Zonotopes

Zonotopes

From Guaranteed State-estimation to Control

Vu Tuan Hieu Le
Cristina Stoica
Teodoro Alamo
Eduardo F. Camacho
Didier Dumur

Series Editor
Francis Castanié

iSTE

WILEY

First published 2013 in Great Britain and the United States by ISTE Ltd and John Wiley & Sons, Inc.

ISTE Ltd
27-37 St George's Road
London SW19 4EU
UK

www.iste.co.uk

John Wiley & Sons, Inc.
111 River Street
Hoboken, NJ 07030
USA

www.wiley.com

Library of Congress Control Number: 2013945045

British Library Cataloguing-in-Publication Data
A CIP record for this book is available from the British Library
ISBN: 978-1-84821-589-4

FSC
www.fsc.org
MIX
Paper from responsible sources
FSC® C013604

Printed and bound in Great Britain by CPI Group (UK) Ltd., Croydon, Surrey CR0 4YY

Table of Contents

Notations

$\mathbb{R}$	Set of real numbers
$\mathbb{R}^+$	Set of strictly positive real numbers
$\mathbb{R}^n$	Set of n-dimensional real vector
$\mathbf{B}^n$	Unitary box in $\mathbb{R}^n$
a	General notation for a scalar
$\mathbf{a}$	General notation for a vector
A	General notation for a matrix
A^T	Transpose of matrix A
A^{-1}	Inverse of matrix A
$det(A)$	Determinant of matrix A
$tr(A)$	Trace of matrix A
$Im(A)$	Image of matrix A
$A \succ 0$	General notation for strictly positive definite matrix A
$A \succeq 0$	General notation for positive definite matrix A
$A \prec 0$	General notation for strictly negative definite matrix A
$A \preceq 0$	General notation for negative definite matrix A
I_n	Identity matrix in $\mathbb{R}^{n \times n}$
0_n	Zeros matrix in $\mathbb{R}^{n \times n}$
$diag(a_1, \ldots, a_n)$	Diagonal matrix of dimension n

$\mathcal{Z} = \mathcal{Z}(\mathbf{p}; H) = \mathbf{p} \oplus H\mathbf{B}^m$	General notation of a m-zonotope
$rs(H)$	Round-sum of matrix H
$\blacksquare\mathcal{Z}$	Approximation of zonotope $\mathcal{Z}$ by a box
$\blacklozenge\mathcal{Z}$	Approximation of zonotope $\mathcal{Z}$ by a parallelotope
$\diamond\mathcal{Z}$	Zonotope inclusion
$\lvert\cdot\rvert$	Absolute value
$\lVert\cdot\rVert_\infty$	Infinity norm
$\lVert\cdot\rVert_P$	P-norm
$\lVert\cdot\rVert_F$	Frobenius norm
$\in$	It belongs to
$\subset$	Subset
$\cap$	Intersection
$\oplus$	Minkowski sum
$\ominus$	Pontryagin difference
$M(\mathcal{S})$	Image of a set $\mathcal{S}$
$d(\mathcal{X}, \mathcal{Y})$	Distance between the sets $\mathcal{X}$ and $\mathcal{Y}$ (also called “normal” distance)
$d_H(\mathcal{X}, \mathcal{Y})$	Hausdorff distance between the sets $\mathcal{X}$ and $\mathcal{Y}$
$\binom{n}{m}$	n combination of m elements
$n!$	Factorial of n
$\mathbf{y}_{k/i}$	The ith row of vector $\mathbf{y}_k$
$conv(\cdot)$	Convex hull
$\omega \sim N(0, Q)$	Random variable ω having zero means, normal distribution and covariance matrix Q

Acronyms

BMI	Bilinear Matrix Inequality
CARIMA	Controlled Auto-Regressive Integrated Moving Average
CRHPC	Constrained Receding Horizon Predictive Control
DMC	Dynamic Matrix Control
EHAC	Extended Horizon Adaptive Control
EPSAC	Extended Prediction Self-Adaptive Control
ESO	Equivalent Single-Output
ESOCE	Equivalent Single-Output with Coupling Effect
EVP	Eigenvalue Problem
GPC	Generalized Predictive Control
LMI	Linear Matrix Inequality
LQ	Linear Quadratic control
LQR	Linear Quadratic Regulator
LTI	Linear Time Invariant
MAC	Model Algorithmic Control
MIMO	Multi-Input Multi-Output
MPC	Model Predictive Control
MPHC	Model Predictive Heuristic Control
MPT	Multi-Parametric Toolbox
MURHAC	Multi-predictor Receding Horizon Adaptive Control

MUSMAR	Multi-step Multivariable Adaptive Control
PAZI	Polytope and Zonotope Intersection
PFC	Predictive Functional Control
PMI	Polynomial Matrix Inequality
QP	Quadratic Programming
SISO	Single-Input Single-Output
SOS	Sum of Squares
SVD	Singular Value Decomposition
TMPC	Tube-based Model Predictive Control
UPC	Unified Predictive Control

Introduction

This book stands at a crossroad between two major axes in automatic control: state estimation and robust control, applied to uncertain discrete-time linear time invariant systems. The goal is to take into account disturbances, measurement noises and constraints in order to build a zonotopic guaranteed state estimation and an output feedback control which can guarantee the feasibility and the stability of the closed-loop system in this specific context. Part of the results proposed in this book were developed and published in the PhD thesis of Hieu Le [LE 12c], under the supervision of the co-authors of this book.

In the literature, there are two main approaches for describing uncertainties, disturbances and noises acting on a dynamic system:

– *Stochastic approach*, which assumes that the disturbances, noises and parameter uncertainties are unknown but the probability distributions are known.

– *Deterministic approach*, which assumes that disturbances, noises and parameter uncertainties are unknown but bounded by some convex sets. The main advantage of the deterministic approach is that disturbances and noises are assumed to be bounded and this is often simpler to verify than the criterion on the probability distribution. This is the main reason why many authors [WIT 68, SCH 68, BER 71] have chosen the deterministic approach to model the disturbances and the noises affecting the system behavior. Based on this remark, the deterministic approach has been chosen in this book to model possible uncertainties, disturbances and/or measurement noises.

Because of the presence of measurement noises, the system state, which is necessary for building the control law, is not always available. In this case, the implementation of a state estimator is necessary. This state estimation problem can be solved by different methods, such as a Luenberger observer [LUE 64], functional observer [MUR 73], moving horizon estimation [GRI 90] and set-membership estimation [WIT 68, SCH 68, BER 71]. Owing to its ability to deal with uncertainties and disturbances, the set-membership estimation method has been chosen in this book. This approach has been applied to the problem of state estimation of uncertain systems since the 1960s [WIT 68, SCH 68, COM 03, ALA 05]. The set-membership estimation allows us to obtain a set containing the real system state that is consistent with the disturbances and measurement noises. With the development of robust control theory, the set-membership estimation technique is shown to be suitable in dealing with unknown but bounded uncertainties, disturbances and measurement noises. If constraints are added to the previous problem, then a predictive control feature should be added. This results in using robust predictive control strategies based on set-membership estimation in order to answer the proposed problem. In particular, zonotopic sets will be used due to their flexibility and low-complexity.

This book builds upon previous results on the zonotopic set-membership state estimation [COM 03, ALA 05] and the output feedback Tube-based Model Predictive Control

[MAY 09]. The aim of the state estimation problem is to obtain a small estimation set which contains the real state. The method proposed by [COM 03] computes a zonotopic outer approximation of the set of states based on a Singular Value Decomposition of a matrix [STR 05], which offers good accuracy of the estimation. In [ALA 05], the authors proposed a method to compute the zonotopic guaranteed state estimation based on two optimization problems. The first solution is based on the minimization of the volume of a zonotope and offers a high-accuracy estimation with a complex computation, while the second solution considers the minimization of the segments of the zonotope and proposes a simple computation but with a deterioration of the estimation accuracy. For these reasons, the goal of this book is to propose a method which permits the improvement of the estimation accuracy while keeping a low complexity level. Moreover, this zonotopic set-membership estimation is proposed to replace the Luenberger observer in the output feedback Tube-based Model Predictive Control [MAY 09]. This association allows us to improve the performance of the closed-loop system as will be shown in the following chapters.

This book is organized as follows:

– *Chapter 1*: the goal of this chapter is to answer the question on how to represent uncertainties, disturbances and noises in the deterministic approach. The chapter starts with a short description of the deterministic approach in which the disturbances and the noises are assumed to be bounded by known convex sets. Afterwards, some basic definitions and operations necessary for manipulating different sets are presented. The next section consists of presenting a list of the most popular families of sets which are used in the literature to bound uncertainties. Because of the advantages of zonotopes, the family of zonotopic sets is further chosen to bound the disturbances and measurement noises.

– *Chapter 2*: in this chapter, an overview of existing estimation techniques is proposed to solve the problem of state estimation for systems subject to unknown but bounded disturbances and measurement noises. Zonotopic-based guaranteed set-membership estimation techniques are further detailed. Two main classes of approaches are addressed: Singular Value Decomposition-based methods [COM 03] and optimization-based methods. The following optimization-based methods are recalled: minimization of the segments of a zonotope [ALA 05] (offering low computation complexity), minimization of the volume of a zonotope [ALA 05] (offering good accuracy of the estimation) and the minimization of the P-radius of a zonotope [LE 11a] (offering a trade-off between the low computation complexity and the accuracy of the state estimation). Moreover, the P-radius criterion allows us to guarantee the non-increasing property of the guaranteed state estimation at each time instant.

– *Chapter 3*: this chapter details the P-radius-based zonotopic set-membership estimation for both Single-Output systems and Multi-Output systems. The Single-Output system solution consists of a zonotopic outer approximation of the intersection between a zonotope and a strip, solved using matrix inequality optimization techniques.

The case of Multi-Output systems leads to two different classes of solutions. The first class is the direct application proposed for the Single-Output systems for each output of the Multi-Output system leading to a conservative result. Several approaches belonging to this first class are developed and compared (the Equivalent Single-Output approach and Equivalent Single-Output with Coupling Effect and Polynomial Matrix Inequality approach). The second class based on the zonotopic outer approximation of the intersection between a polytope and a zonotope allows us to improve the accuracy of the estimation while considering all the output measurements at the same time.

– *Chapter 4*: the problem of robust predictive control is discussed in this chapter, in the context of zonotopic set-membership estimation. Based on the zonotopic set-membership estimation discussed in Chapter 3, a feedback predictive control based on a tube of trajectories is proposed for the case of linear discrete-time invariant systems with bounded disturbances and measurement noises, subject to constraints. This chapter also proposes an application of the developed approaches to control a magnetic levitation system. The first step consists of describing and modeling this nonlinear unstable continuous-time system subject to bounded disturbances, measurement noises and constraints. The proposed model is linearized around an equilibrium point and discretized for a given sampling time. Based on this model, the Tube-based Model Predictive Control associated with the zonotopic set-membership estimation is used to stabilize this system around the equilibrium point.

– *Conclusion and Perspectives*: the final chapter summarizes the work developed in this book and proposes several directions for future developments.

Chapter 1

Uncertainty Representation Based on Set Theory

Real systems are often complex due to several factors: the system's nature (e.g. mechanical, electrical and chemical systems), interactions between its different components (e.g. multivariable systems), and its different behavior in a dynamic environment (e.g. influence of disturbances, noises and uncertainties). All these aspects have to be considered when modeling a given system, sometimes leading to a complicated model. In the context of control systems, a mathematical model is frequently used to describe the system behavior. On the one hand, the accuracy of the mathematical model is important to analyze and design control strategies for the considered system; on the other hand, in the context of industrial applications, it is suitable to use unsophisticated controllers designed using a simple model. In this context, a trade-off must be found: the system model should be simple but precise enough to characterize the dynamical behavior of the original system. Thus, the simple/simplified mathematical model cannot represent the real system exactly due to a lack of knowledge of, or unreliable information about, the system. To validate this model, some uncertainties can be added to the mathematical model. Frequently, perturbations influencing the real system have to be taken into account in the mathematical model in order to ensure a similar behavior of the real system and the mathematical model. The importance of uncertainties in system design has been discussed by many authors (the interested reader can refer to [MAY 79, AUG 06, AYY 06] and the references therein). In the literature, there are two ways to represent uncertainties: the statistical (or stochastic) approach and the deterministic approach. An overview of these two directions is provided in the following.

In the *stochastic approach*, the uncertainty is modeled by a random process with a known statistical property. This technique is widely used in various domains (e.g. economics [BAT 08], biology [ULL 11] and engineering [MAY 79]), especially when estimates of the probability distribution of the uncertain parameters are available. But in many applications, there are situations when the probability distribution of the uncertain parameters is not known and only bounds of the uncertain domain can be fixed. In this case, the probabilistic assumptions on the uncertainty are no longer valid, making this method unsuitable for modeling the uncertainties.

In the *deterministic approach*, the uncertainty is assumed to belong to a set: a classical (crisp) set [1] or a fuzzy set [2]. In the literature, different families of classical sets are used depending on their properties. Usually, the accuracy and the

1 A classical set is a set wherein the degree of membership of any object to the set is either 0 or 1.

2 A fuzzy set is a set wherein the degree of membership of any object to the set is between 0 and 1.

complexity of the uncertainties' representation are inversely proportional, depending on the particular problem related to the choice of a suitable geometric form. In the following sections, some popular families of sets are presented with their advantages and weaknesses. Note that in this book only convex classical sets are considered due to the important role of convexity in the optimization theory [BER 03].

1.1. Basic set definitions: advantages and weaknesses

Before presenting the most well-known families of sets, some basic set definitions and operations used in this book are introduced.

DEFINITION 1.1.– *A set $\mathcal{S} \subset \mathbb{R}^n$ is called a* convex set *if for any $\boldsymbol{x}_1, \boldsymbol{x}_2, \ldots, \boldsymbol{x}_k \in \mathcal{S}$ and any $\alpha_1, \alpha_2, \ldots, \alpha_k \in \mathbb{R}^+$ such that $\sum_{i=1}^{k} \alpha_i = 1$, then the element $\sum_{i=1}^{k} \alpha_i \boldsymbol{x}_i$ is in $\mathcal{S}$.*

DEFINITION 1.2.– *A* convex hull *of a given set $\mathcal{S}$, denoted $conv(\mathcal{S})$, is the smallest convex set containing $\mathcal{S}$.*

DEFINITION 1.3.– *A set $\mathcal{S} \subset \mathbb{R}^n$ is called a* C-set *if $\mathcal{S}$ is compact, convex and contains the origin. This is a* proper set *if its interior is not empty.*

DEFINITION 1.4.– *The* inclusion *operator between two sets is defined by $\mathcal{X} \subseteq \mathcal{Y}$, if and only if $\forall x \in \mathcal{X}$, then $x \in \mathcal{Y}$. This means that $\mathcal{X}$ is a subset of $\mathcal{Y}$.*

DEFINITION 1.5.– *The* intersection *operator of two sets $\mathcal{X}$ and $\mathcal{Y}$ is defined by $\mathcal{X} \cap \mathcal{Y} = \{x : x \in \mathcal{X} \text{ and } x \in \mathcal{Y}\}$.*

DEFINITION 1.6.– *The* image *of a set S under a map (projection) $\mathcal{M}$ is the set $\mathcal{M}(\mathcal{S}) = \{y : y = \mathcal{M}(x), x \in \mathcal{S}\}$.*

DEFINITION 1.7.– *The* Minkowski sum *of two sets $\mathcal{X}$ and $\mathcal{Y}$ is defined by* $\mathcal{X} \oplus \mathcal{Y} = \{x + y : x \in \mathcal{X}, y \in \mathcal{Y}\}$.

DEFINITION 1.8.– *The* Pontryagin difference *of two sets $\mathcal{X}$ and $\mathcal{Y}$ is defined by* $\mathcal{X} \ominus \mathcal{Y} = \{z : z + y \in \mathcal{X}, \forall y \in \mathcal{Y}\}$.

DEFINITION 1.9.– *Let $\mathcal{X}$ and $\mathcal{Y}$ be two non-empty sets. The* distance *between these two sets $\mathcal{X}$ and $\mathcal{Y}$ is defined by* $d(\mathcal{X}, \mathcal{Y}) = min\{d(x, y) : x \in \mathcal{X}, y \in \mathcal{Y}\}$.

DEFINITION 1.10.– *Let $\mathcal{X}$ and $\mathcal{Y}$ be two non-empty sets. The* Hausdorff distance *of these two sets $\mathcal{X}$ and $\mathcal{Y}$ is defined by* $d_H(\mathcal{X}, \mathcal{Y}) = max\{\bar{d}_H(\mathcal{X}, \mathcal{Y}), \bar{d}_H(\mathcal{Y}, \mathcal{X})\}$, *with* $\bar{d}_H(\mathcal{X}, \mathcal{Y}) = \max_{x \in \mathcal{X}} \min_{y \in \mathcal{Y}} d(x, y)$.

The Hausdorff distance allows characterizing the quality of the approximation of $\mathcal{X}$ by $\mathcal{Y}$ [HUN 93]. If $\mathcal{X}$ and $\mathcal{Y}$ have the same closure [3], then the Hausdorff distance is equal to 0. Figure 1.1 illustrates the difference between the "normal" distance (Definition 1.9) which is equal to 0 and the Hausdorff distance $d_H(\mathcal{X}, \mathcal{Y})$ between the two sets $\mathcal{X}$ and $\mathcal{Y}$.

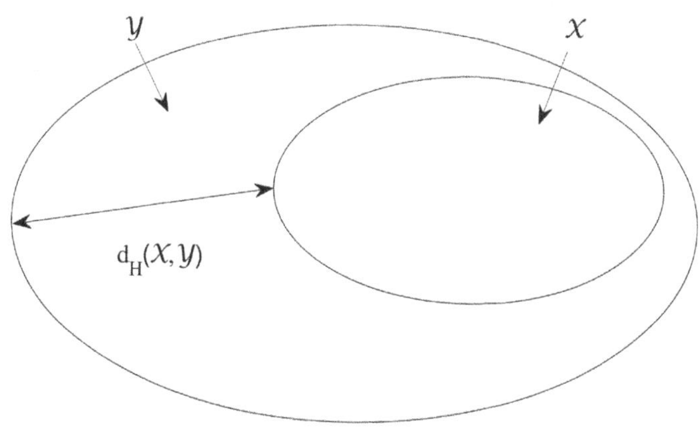

Figure 1.1. *Illustration of the Hausdorff distance between two sets*

3 The closure of a set $\mathcal{S}$ is defined by the union of $\mathcal{S}$ and its boundary $\partial\mathcal{S}$.

1.1.1. *Interval set*

A very simple way to define uncertainties is by using the *interval* notion. This is based on the idea of enclosing numerical errors into an interval. In many cases, obtaining the probability of occurrence of different uncertainties is not possible. Therefore, it can be easier and thus suitable to bound the uncertainties by intervals. Moreover, the interval analysis allows us to simplify most of the standard operations [MOO 66, HAN 65, JAU 01]. This approach is developed in many domains (e.g. identification, diagnosis and estimation), especially when a short computation time is required.

DEFINITION 1.11.– *An* interval $\boldsymbol{I} = [a, b] \subset \mathbb{R}$ *is defined by the set* $\{x \in \mathbb{R} : a \leq x \leq b\}$.

DEFINITION 1.12.– *The* center *and the* radius *of an interval* $\boldsymbol{I} = [a, b]$ *are, respectively, defined by* $mid(\boldsymbol{I}) = \frac{a+b}{2}$ *and* $rad(\boldsymbol{I}) = \frac{b-a}{2}$.

DEFINITION 1.13.– *The* unitary interval *is denoted by* $\boldsymbol{B} = [-1, 1]$.

DEFINITION 1.14.– *The* set of real compact intervals $[a, b]$, *where* $a, b \in \mathbb{R}$ *and* $a \leq b$, *is denoted by* $\mathbb{I}$.

DEFINITION 1.15.– *A* box $([a_1, b_1], \ldots, [a_n, b_n])^T$ *is an interval vector.*

DEFINITION 1.16.– *A* unitary box *in* $\mathbb{R}^n$, *denoted by* $\boldsymbol{B}^n$, *is a box composed of* n *unitary intervals.*

Consider the intervals $[\underline{x}, \bar{x}]$ and $[\underline{y}, \bar{y}]$. An operation $\circ$ between these two intervals can be formalized as:

$$[\underline{x}, \bar{x}] \circ [\underline{y}, \bar{y}] = \{x \circ y : x \in [\underline{x}, \bar{x}], y \in [\underline{y}, \bar{y}]\} \qquad [1.1]$$

The four basic operations [MOO 66] with intervals are the following:

1) $[\underline{x}, \bar{x}] + [\underline{y}, \bar{y}] = [\underline{x} + \underline{y}, \bar{x} + \bar{y}]$

2) $[\underline{x}, \bar{x}] - [\underline{y}, \bar{y}] = [\underline{x} - \bar{y}, \bar{x} - \underline{y}]$

3) $[\underline{x}, \bar{x}] * [\underline{y}, \bar{y}] = [\min(\underline{x} \cdot \underline{y}, \underline{x} \cdot \bar{y}, \bar{x} \cdot \underline{y}, \bar{x} \cdot \bar{y}), \max(\underline{x} \cdot \underline{y}, \underline{x} \cdot \bar{y}, \bar{x} \cdot \underline{y}, \bar{x} \cdot \bar{y})]$

4) $[\underline{x}, \bar{x}] / [\underline{y}, \bar{y}] = [\underline{x}, \bar{x}] * [1/\bar{y}, 1/\underline{y}]$, if $0 \notin [\underline{y}, \bar{y}]$

Despite the simplicity of interval analysis, a drawback of this approach is that the computation results can sometimes be conservative due to the *dependency effect*[4] and the *wrapping effect*[5] [MOO 66, KÜH 98b, JAU 01]. These two effects are further analyzed through two examples.

EXAMPLE 1.1.– DEPENDENCY EFFECT – Consider two functions $f_1(x, y) = x - y$ and $f_2(x) = x - x$, with the variables $x, y \in [-1, 1]$. Using the interval analysis, the value domain of f_1 and f_2 is the same $[-1, 1] - [-1, 1] = [-2, 2]$, even if the real value domain of f_2 is 0. This problem, called the "dependency effect", relies on the fact that the occurrence of the same variable x in the function f_2 is independently considered and can lead to an important over-approximation of the result.

EXAMPLE 1.2.– WRAPPING EFFECT – Consider a function $f(x, y) = \begin{bmatrix} f_1(x, y) \\ f_2(x, y) \end{bmatrix} = A \cdot \begin{bmatrix} x \\ y \end{bmatrix}$, with $x, y \in [-1, 1]$ and $A = \begin{bmatrix} 0 & -0.5 \\ 1 & 1 \end{bmatrix}$. Using the interval analysis leads to $f(x, y) = \begin{bmatrix} f_1(x, y) \\ f_2(x, y) \end{bmatrix} = \begin{bmatrix} -0.5y \\ x + y \end{bmatrix}$, with the value domains $f_1 \in [-0.5, 0.5]$ and $f_2 \in [-2, 2]$, represented in dark gray in

4 The dependency effect appears when a variable occurs more than one time in a function and each occurrence is considered independently, increasing the resulting interval.

5 The wrapping effect appears when the domain representation grows due to the over-estimation at each sampling time.

Figure 1.2. The vertices of the exact solution of f (see Figure 1.2) $\left\{\begin{bmatrix}-0.5\\2\end{bmatrix},\begin{bmatrix}-0.5\\0\end{bmatrix},\begin{bmatrix}0.5\\-2\end{bmatrix},\begin{bmatrix}0.5\\0\end{bmatrix}\right\}$ are obtained by computing the product of the matrix A with each vertex of the unitary box containing x and y (e.g. $[-0.5, 0]^T = A \cdot [-1, 1]^T$). Comparing these solutions, an important over-approximation of the interval analysis solution can be observed. If this operation is repeated several times, the difference between the exact solution and the solution of the interval analysis is more and more important. This problem is called the "wrapping effect".

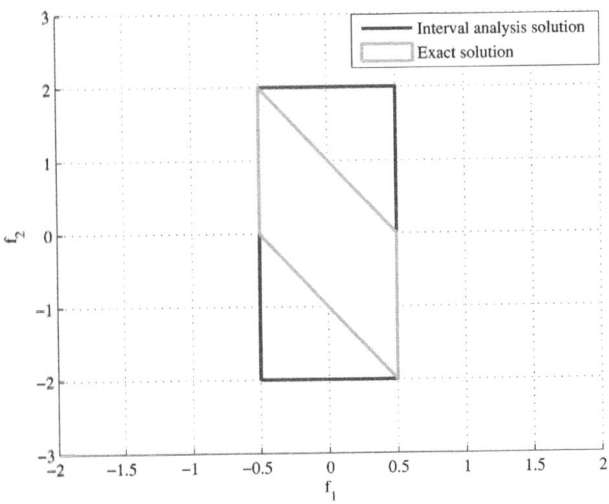

Figure 1.2. *Illustration of the wrapping effect in interval analysis*

1.1.2. *Ellipsoidal set*

Another way to represent bounded uncertainties is based on the use of bounded ellipsoids. Moreover, due to its low complexity, the *ellipsoidal* set is widely used in a large class of applications in automatic control [SCH 68, WIT 68,

BER 71, CHE 81, MIL 96, KUR 96, DUR 01, POL 04]. Some basic properties of ellipsoids are discussed in the following.

DEFINITION 1.17.– *Given a vector* $c \in \mathbb{R}^n$ *and a symmetric positive definite matrix* $P = P^T \succ 0$*, an* ellipsoid $\mathcal{E}(\mathbf{c}, P)$ *is defined as follows:*

$$\mathcal{E}(\mathbf{c}, P) = \{\boldsymbol{x} \in \mathbb{R}^n : (\boldsymbol{x} - \mathbf{c})^T P^{-1}(\boldsymbol{x} - \mathbf{c}) \leq 1\} \qquad [1.2]$$

The vector $\mathbf{c}$ is called the *center* of the ellipsoid $\mathcal{E}(\mathbf{c}, P)$ and the matrix P is called the *shape matrix* of the ellipsoid $\mathcal{E}(\mathbf{c}, P)$.

EXAMPLE 1.3.– An ellipsoid $\mathcal{E}(\mathbf{c}, P)$ with $\mathbf{c} = \begin{bmatrix} 0 \\ 0 \end{bmatrix}$ and $P = \begin{bmatrix} 1 & 1 \\ 1 & 4 \end{bmatrix}$ is illustrated in Figure 1.3.

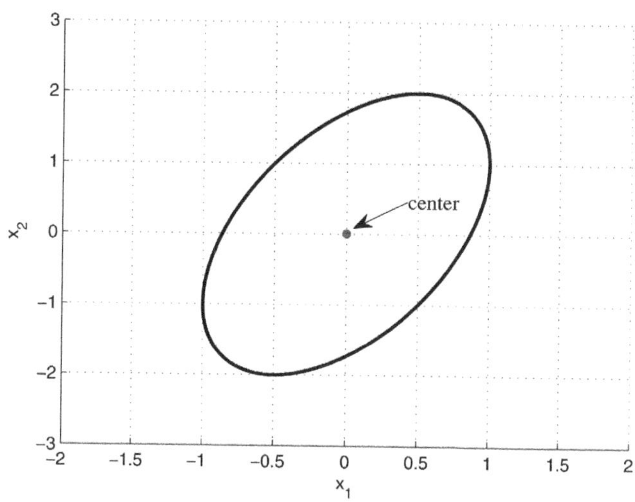

Figure 1.3. *Illustration of a bounded ellipsoid in* $\mathbb{R}^2$

PROPERTY 1.1.– Consider an ellipsoid $\mathcal{E}(\mathbf{c}, P) \subseteq \mathbb{R}^n$, a matrix $A \in \mathbb{R}^{n \times n}$ and a vector $\mathbf{b} \in \mathbb{R}^n$. An *affine transformation* of this ellipsoid is also an ellipsoid $A\mathcal{E}(\mathbf{c}, P) + \mathbf{b} = \mathcal{E}(A\mathbf{c} + \mathbf{b}, APA^T)$.

The interested reader can find concrete studies on ellipsoids and their operations in [BOY 94] and [KUR 96] and can easily use the Ellipsoidal Toolbox [KUR 06] to implement different operations of ellipsoidal calculus within a MATLAB® environment.

Despite the simple representation of ellipsoids, there are still some drawbacks that can be a source of conservatism. The main inconvenience is that ellipsoids are not closed under some operations [6]. Furthermore, ellipsoids allow representing bounded uncertainties which are necessarily coupled [LAL 08] due to the low flexibility of the ellipsoidal shape compared to polyhedral sets which are presented in the following section.

1.1.3. *Polyhedral set*

The *polyhedral* set is one of the most popular geometrical forms used in control systems and optimization. A polyhedral set in a finite-dimensional Euclidean space is the intersection of a finite amount of closed half-spaces [ZIE 95]. A bounded polyhedral is denoted as a *polytope*. Because of its flexibility, polytopes offer a good approximation of any convex set [LAY 82, BRO 08, SCI 11]. In addition, polytopes are closed under the basic operations [7] unlike ellipsoids. Moreover, its dual representations (half-space representation and vertex representation) allow us to choose the appropriate form for a particular problem. The main weakness of polytopes is due to

---

6 For example, the Minkowski sum or the intersection of two ellipsoids is generally not an ellipsoid. In order to preserve the simplicity offered by ellipsoids, ellipsoidal outer approximations of the obtained sets can be made, leading to a trade-off between accuracy of the representation and simplicity of the computation.

7 For example, the Minkowski sum or the intersection of two convex polytopes is generally a polytope.

their complexity related to the number of vertices, but not fixed by the space dimension. Therefore, even if a polytope can approximate any convex set well, the complexity can quickly increase with the number of vertices even in a low space dimension. In order to formalize the notations, the main definitions of polytopes are summarized below.

DEFINITION 1.18.– HALF-SPACE REPRESENTATION – *A* polyhedral set $\mathcal{P} \in \mathbb{R}^n$ *can be defined as the intersection of a finite number of half-spaces:* $\mathcal{P} = \{\boldsymbol{x} \in \mathbb{R}^n : H\boldsymbol{x} \leq \boldsymbol{k}\}$, *with* $H \in \mathbb{R}^{m \times n}$, $\boldsymbol{k} \in \mathbb{R}^m$. *If* $\mathcal{P}$ *is bounded, then* $\mathcal{P}$ *is a* polytope.

EXAMPLE 1.4.– A polytope $\mathcal{P}$ in the half-space representation with:

$$H = \begin{bmatrix} -1 & 1\,1 & -1 \\ 1 & -1\,1 & -4 \end{bmatrix}^T, \mathbf{k} = \begin{bmatrix} 1\,1\,1\,4 \end{bmatrix}^T$$

is shown in Figure 1.4. H_i and k_i denote the ith row of matrix H and the ith element of vector $\mathbf{k}$, respectively.

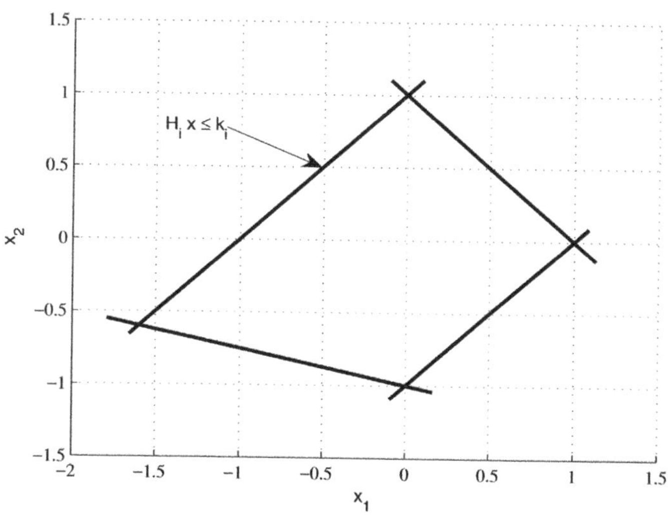

Figure 1.4. *H-representation of a polytope*

DEFINITION 1.19.– VERTEX REPRESENTATION – *For a finite set of points $\mathcal{V} \subset \mathbb{R}^n$, with $\mathcal{V} = \{\boldsymbol{v}_1, \boldsymbol{v}_2, \ldots, \boldsymbol{v}_m\}$, a* polytope $\mathcal{P}$ *can be defined as the convex hull of the set $\mathcal{V}$:*

$$\mathcal{P} = conv(\mathcal{V}) = \Big\{\alpha_1 \mathbf{v}_1 + \alpha_2 \mathbf{v}_2 + \ldots + \alpha_m \mathbf{v}_m : \alpha_i \in \mathbb{R}^+, \sum_{i=1}^{m} \alpha_i = 1, \mathbf{v}_i \in \mathbb{R}^n \Big\}$$

EXAMPLE 1.5.– Consider the set of vertices $\mathcal{V} = \left\{ \begin{bmatrix} 0 \\ 1 \end{bmatrix}, \begin{bmatrix} 1 \\ 0 \end{bmatrix}, \begin{bmatrix} 0 \\ -1 \end{bmatrix}, \begin{bmatrix} -1.6 \\ -0.6 \end{bmatrix} \right\}$. Figure 1.5 shows the V-representation of a polytope based on the set $\mathcal{V}$.

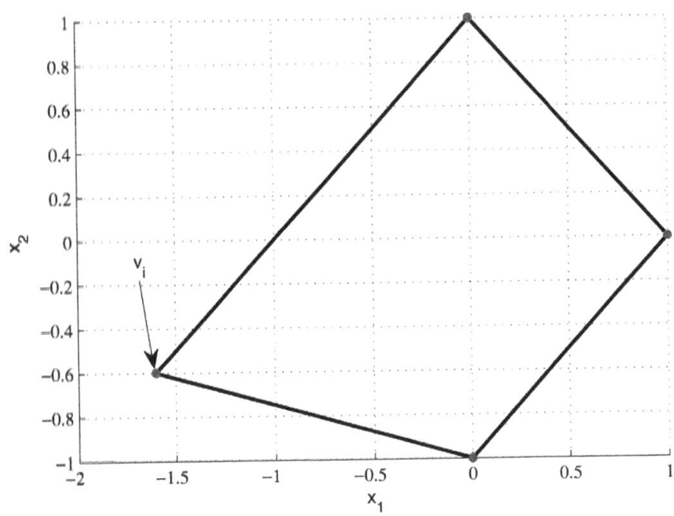

Figure 1.5. *V-representation of a polytope. For a color version of this figure, see www.iste.co.uk/stoica/zonotopes.zip*

Theorem 1.1 [ZIE 95] shows the equivalence of these two definitions of a polytope (i.e. H-polytope and V-polytope) allowing us to choose a suitable representation for a particular problem. For example, the proof that the polytopic set is closed under the Minkowski sum is trivial via the V-representation but can become complicated via the H-representation.

THEOREM 1.1.– EQUIVALENCE OF H-REPRESENTATION AND V-REPRESENTATION [ZIE 95] – A set $\mathcal{P} \subset \mathbb{R}^n$ is the convex hull of a finite point set (a V-polytope) if and only if it is a bounded intersection of half-spaces (a H-polytope).

This theorem shows that the H-polytope can be transformed into the V-polytope and vice versa. In the literature, this problem is well known as the *vertex enumeration problem* for the transformation of a V-polytope to a H-polytope and the *facet enumeration problem* for the transformation of a H-polytope to its equivalent V-polytope. There exist algorithms to solve these transformation problems, but they are time consuming (e.g. [DAN 72] and [FUK 99]). The interested reader can find more details on polytopes in [ZIE 95] and [BLA 07] and the references therein. An example of the same polytope defined by H-representation and V-representation is illustrated in Figures 1.4 and 1.5.

Even if polytopes can approximate any convex set well, their complexity increases with its number of vertices. This can lead to some limitations for the direct application of the methods based on polytopes. In the following section, a geometrical form which offers a good trade-off between complexity and flexibility is presented.

1.1.4. *Zonotopic set*

In recent years, zonotopes are used more and more to represent uncertainties due to the flexibility, the reduced complexity and specially the efficient computation of linear transformations and Minkowski sums. Zonotopes are a special class of convex symmetric polytopes. Thus, the properties developed for convex polytopes can also be applied to zonotopes. In this way, zonotopes can be represented as H-polytopes or V-polytopes. There also exist different representations specific to zonotopes, which are presented below.

DEFINITION 1.20.– GENERATOR REPRESENTATION – *Given a vector* $\boldsymbol{p} \in \mathbb{R}^n$ *and a set of vectors* $\mathcal{G} = \{\boldsymbol{g}_1, \boldsymbol{g}_2, \cdots, \boldsymbol{g}_m\} \subset \mathbb{R}^n$*,* $m \geq n$*, a zonotope* $\mathcal{Z}$ *of order* m *(also called a* m*-zonotope) is defined as follows:*

$$\mathcal{Z} = (\boldsymbol{p}; \boldsymbol{g}_1, \boldsymbol{g}_2, \cdots, \boldsymbol{g}_m)$$

$$= \{\boldsymbol{x} \in \mathbb{R}^n : \boldsymbol{x} = \boldsymbol{p} + \sum_{i=1}^{m} \alpha_i \boldsymbol{g}_i; -1 \leq \alpha_i \leq 1\} \quad [1.3]$$

The vector $\mathbf{p}$ is called the *center* of the zonotope $\mathcal{Z}$. The vectors $\mathbf{g}_1, \cdots, \mathbf{g}_m$ are called the *generators* of $\mathcal{Z}$. The *order* of a zonotope is defined by the number of its generators (m in this case). A m-zonotope $\mathcal{Z} \subset \mathbb{R}^n$ with m generators and $m < n$ is called a *degenerated* m*-zonotope*.

This definition is equivalent to the definition of zonotopes obtained by the Minskowski sum of a finite number of line segments defined by $\mathbf{g}_i\mathbf{B}^1$:

$$\mathcal{Z} = (\mathbf{p}; \mathbf{g}_1, \mathbf{g}_2, \cdots, \mathbf{g}_m) = \mathbf{p} \oplus \mathbf{g}_1\mathbf{B}^1 \oplus \ldots \oplus \mathbf{g}_m\mathbf{B}^1 \quad [1.4]$$

EXAMPLE 1.6.– Consider the following zonotope of third order in $\mathbb{R}^2$ with $\mathbf{p} = \begin{bmatrix} 0 \\ 0 \end{bmatrix}$ and its generators $\mathbf{g}_1 = \begin{bmatrix} 1 \\ 3 \end{bmatrix}$, $\mathbf{g}_2 = \begin{bmatrix} 2 \\ 2 \end{bmatrix}$, $\mathbf{g}_3 = \begin{bmatrix} 3 \\ 1 \end{bmatrix}$. This 3-zonotope is illustrated in Figure 1.6.

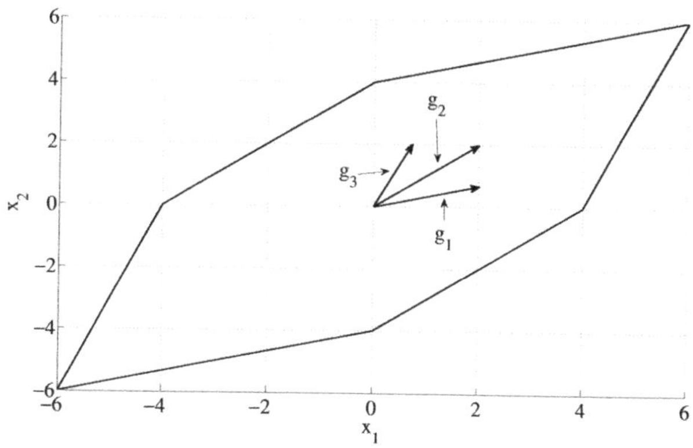

Figure 1.6. *3-zonotope and its generators in $\mathbb{R}^2$*

EXAMPLE 1.7.– Figure 1.7 represents a sixth-order *centered zonotope*[8] in $\mathbb{R}^3$ with:

$$\mathbf{p} = \begin{bmatrix} 0 \\ 0 \\ 0 \end{bmatrix}, \mathbf{g}_1 = \begin{bmatrix} 1 \\ 1 \\ 0 \end{bmatrix}, \mathbf{g}_2 = \begin{bmatrix} 1 \\ -1 \\ 0 \end{bmatrix}, \mathbf{g}_3 = \begin{bmatrix} 1 \\ 0 \\ 1 \end{bmatrix}, \mathbf{g}_4 = \begin{bmatrix} 1 \\ 0 \\ -1 \end{bmatrix},$$
$$\mathbf{g}_5 = \begin{bmatrix} 0 \\ 1 \\ 1 \end{bmatrix}, \mathbf{g}_6 = \begin{bmatrix} 0 \\ 1 \\ -1 \end{bmatrix}.$$

From these two examples, it can be observed that the complexity of zonotopes depends on the number of generators and the dimension of the space. The complexity grows quickly with the number of generators: the number of vertices of the zonotope is 6 in Figure 1.6 and 24 in Figure 1.7. When the number of generators is increased, the space dimension is also increased in Figure 1.7 compared to Figure 1.6.

Another definition of zonotopes that is more convenient for the approach considered in this book is based on the fact that

8 A centered zonotope is a zonotope whose center is the origin (i.e. $\mathbf{p} = 0_n$ in equation [1.3]).

a m-zonotope in $\mathbb{R}^n$ can be defined as the affine image of a m-dimensional hypercube in $\mathbb{R}^n$.

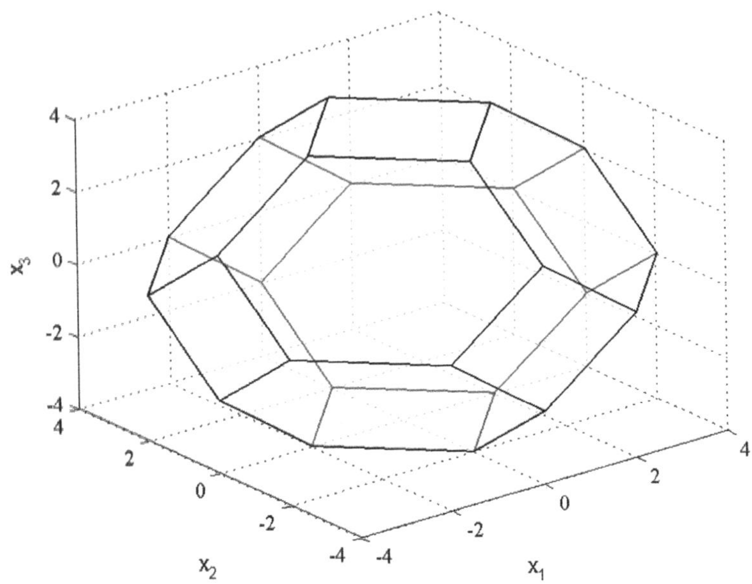

Figure 1.7. *6-zonotope in* $\mathbb{R}^3$

DEFINITION 1.21.– HYPERCUBE AFFINE PROJECTION – *A m-zonotope in $\mathbb{R}^n$ ($m \geq n$) is the translation by the center $\boldsymbol{p} \in \mathbb{R}^n$ of the image of a unitary hypercube of dimension m in $\mathbb{R}^n$ under a linear transformation. Given a matrix $H \in \mathbb{R}^{n \times m}$ representing the linear transformation, the zonotope $\mathcal{Z}$ is defined by:*

$$\mathcal{Z} = (\boldsymbol{p}; H) = \boldsymbol{p} \oplus H\boldsymbol{B}^m \qquad [1.5]$$

Considering the matrix $H = \left[\mathbf{g}_1\, \mathbf{g}_2 \cdots \mathbf{g}_m\right]$ proves the equivalence of the two proposed definitions. From now on, to simplify the chapter, the zonotope $\mathcal{Z}$ is described by $\mathcal{Z}(\mathbf{p}; H)$.

EXAMPLE 1.8.– Consider the zonotope $\mathcal{Z}(\mathbf{p}; H)$ with $\mathbf{p} = \begin{bmatrix} 0 \\ 0 \end{bmatrix}$ and $H = \begin{bmatrix} 1 & 2 & 3 \\ 3 & 2 & 1 \end{bmatrix}$. Using the hypercube affine projection, this zonotope can be constructed as the image of a hypercube in $\mathbb{R}^3$ (with its eight vertices $\begin{bmatrix} 1 \\ 1 \\ 1 \end{bmatrix}, \begin{bmatrix} -1 \\ 1 \\ 1 \end{bmatrix}, \begin{bmatrix} 1 \\ -1 \\ 1 \end{bmatrix}, \begin{bmatrix} 1 \\ 1 \\ -1 \end{bmatrix}, \begin{bmatrix} -1 \\ -1 \\ 1 \end{bmatrix}, \begin{bmatrix} -1 \\ 1 \\ -1 \end{bmatrix}, \begin{bmatrix} 1 \\ -1 \\ -1 \end{bmatrix}, \begin{bmatrix} -1 \\ -1 \\ -1 \end{bmatrix}$) under the projection H in $\mathbb{R}^2$ (see Figure 1.8). The reader can note that the same zonotope is obtained via the hypercube affine projection (Figure 1.8) and via the generator representation (Figure 1.6).

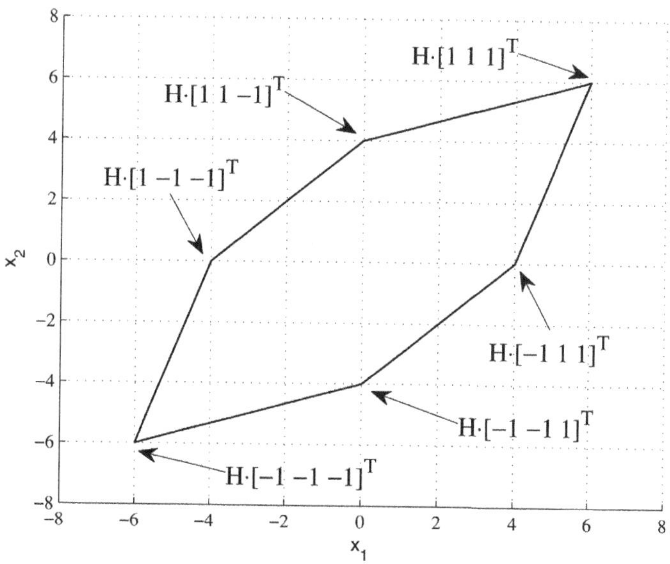

Figure 1.8. *3-zonotope and its vertices in* $\mathbb{R}^2$

The conversion between these particular representations of zonotopes is studied by many authors

[GRI 93, SEY 94, FUK 04, SCH 05b, ALT 10]. As zonotopes are special cases of polytopes, the generator representation and the hypercube affine projection of a zonotope can be converted to the V-representation and also to the H-representation. These conversions are related to the Minkowski sum of two polytopes because the generator representation is equivalent to the Minkowski sum of a finite number of line segments, which is a polytope. In addition, the generator representation illustrates a significant advantage of zonotopes: a complex geometrical form represented via a simple matrix and a vector (that can be zero for zonotopes centered to the origin). For example, the centered zonotope from Figure 1.7 with 24 vertices in $\mathbb{R}^3$ is represented by a matrix $H \in \mathbb{R}^{3\times 6}$. This leads us to simplify the mentioned basic set operations by simple matrix computation as presented in the following section.

1.2. Main properties of zonotopes

This section proposes an overview of the main properties of zonotopes that will be used throughout this book. Some examples are also considered in order to make the book easier to comprehend.

PROPERTY 1.2.– PARTICULAR ZONOTOPES – *Given a centered zonotope* $\mathcal{Z} = H\boldsymbol{B}^m \in \mathbb{R}^n$. Because of the properties of the matrix H, some particular forms of zonotopes can be obtained. If H is the identity matrix, then $\mathcal{Z}$ is the *unit box*. If H is diagonal, orthogonal or invertible, then $\mathcal{Z}$ is a *box*, a *hypercube* or a *parallelotope* [9], respectively.

9 A parallelotope is a special zonotope whose number of generators is equal to the dimension of the space.

PROPERTY 1.3.– GENERATOR PERMUTATION – The permutation of the matrix columns in the generator representation of a zonotope does not modify the zonotope.

PROOF – This property results from the commutativity of the Minkowski sum of several line segments. □

PROPERTY 1.4.– SUM OF TWO ZONOTOPES – Given two zonotopes $\mathcal{Z}_1 = \mathbf{p}_1 \oplus H_1\mathbf{B}^{m_1} \in \mathbb{R}^n$ and $\mathcal{Z}_2 = \mathbf{p}_2 \oplus H_2\mathbf{B}^{m_2} \in \mathbb{R}^n$, the Minkowski sum of two zonotopes is also a zonotope defined by $\mathcal{Z} = \mathcal{Z}_1 \oplus \mathcal{Z}_2 = (\mathbf{p}_1 + \mathbf{p}_2) \oplus \begin{bmatrix} H_1 & H_2 \end{bmatrix} \mathbf{B}^{m_1+m_2}$.

PROOF – From the definition of the Minkowski sum, it results in $\mathcal{Z}_1 \oplus \mathcal{Z}_2 = \{\mathbf{p}_1 + \mathbf{p}_2 + H_1\mathbf{z}_1 + H_2\mathbf{z}_2 : \mathbf{z}_1 \in \mathbf{B}^{m_1}, \mathbf{z}_2 \in \mathbf{B}^{m_2}\}$, which can be further rewritten in a matrix form as: $\mathcal{Z}_1 \oplus \mathcal{Z}_2 = \{\mathbf{p}_1 + \mathbf{p}_2 + \begin{bmatrix} H_1 & H_2 \end{bmatrix} \cdot \begin{bmatrix} \mathbf{z}_1 \\ \mathbf{z}_2 \end{bmatrix} : \begin{bmatrix} \mathbf{z}_1 \\ \mathbf{z}_2 \end{bmatrix} \in \mathbf{B}^{m_1+m_2}\} = (\mathbf{p}_1 + \mathbf{p}_2) \oplus \begin{bmatrix} H_1 & H_2 \end{bmatrix} \mathbf{B}^{m_1+m_2} = \mathcal{Z}$. □

PROPERTY 1.5.– LINEAR IMAGE OF A ZONOTOPE – *The image of a zonotope* $\mathcal{Z} = \boldsymbol{p} \oplus H\boldsymbol{B}^m \in \mathbb{R}^n$ *by a linear mapping* K *can be computed by a standard matrix product* $K \cdot \mathcal{Z} = (K \cdot \boldsymbol{p}) \oplus (K \cdot H)\boldsymbol{B}^m$.

PROOF – Using matrix multiplication leads to $K\mathcal{Z} = \{K(\mathbf{p} + H\mathbf{z}) : \mathbf{z} \in \mathbf{B}^m\} = \{K\mathbf{p} + KH\mathbf{z} : \mathbf{z} \in \mathbf{B}^m\} = (K\mathbf{p}) \oplus (KH)\mathbf{B}^m$. □

The basic operations with zonotopes can increase the size of the obtained zonotopes. Thus, an interesting issue concerns the techniques that reduce the complexity of zonotopes, allowing us to limit the number of generators of a zonotope. For example, if the problem of *reachable set* [10] is addressed using zonotopes, the complexity of this zonotope increases at each sample time due to the Minkowski sum operation. This

10 This is the problem of computing all states visited by trajectories of a system starting from any $x_0 \in X_0$.

part focuses on over-approximations of a high-order zonotope leading to compute a reduced order zonotope enclosing the initial zonotope.

PROPOSITION 1.1.– COMPLEXITY REDUCTION OF ZONOTOPES VIA INTERVAL HULL METHOD – Considering a zonotope $\mathcal{Z} = \mathbf{p} \oplus H\mathbf{B}^m \in \mathbb{R}^n$, the smallest box containing this zonotope is computed by:

$$\blacksquare\mathcal{Z} = \mathbf{p} \oplus rs(H)\mathbf{B}^n \qquad [1.6]$$

with $rs(H)$ a diagonal matrix such that $rs(H)_{ii} = \sum_{j=1}^{m} |H_{ij}|$, $i = 1, \ldots, n$.

PROOF.– As a box is an axis-aligned set, the over-approximation of a zonotope by a box can be made by considering its extreme points in each direction. The extreme point in the direction i can be computed by $p_i + \sum_{j=1}^{m} |H_{ij}|$. All extreme points in all n directions are similarly computed and the smallest box containing the zonotope $\mathcal{Z}$ is obtained as $\blacksquare\mathcal{Z} = \mathbf{p} \oplus rs(H)\mathbf{B}^n$. □

This proposition provides a simple and fast over-approximation of a zonotope by a box. The result has a minimal complexity which is given by the dimension of the space. However, the result obtained with this proposition is conservative because the form of the initial zonotope is lost.

An example is proposed in the following in order to better illustrate this proposition.

EXAMPLE 1.9.– Consider a centered zonotope $\mathcal{Z} = H\mathbf{B}^3 \in \mathbb{R}^2$, with $H = \begin{bmatrix} 1 & 2 & 3 \\ 3 & 2 & 1 \end{bmatrix}$. Applying the interval hull approximation leads to a box (in Figure 1.9) containing the original zonotope (hexagon in Figure 1.9).

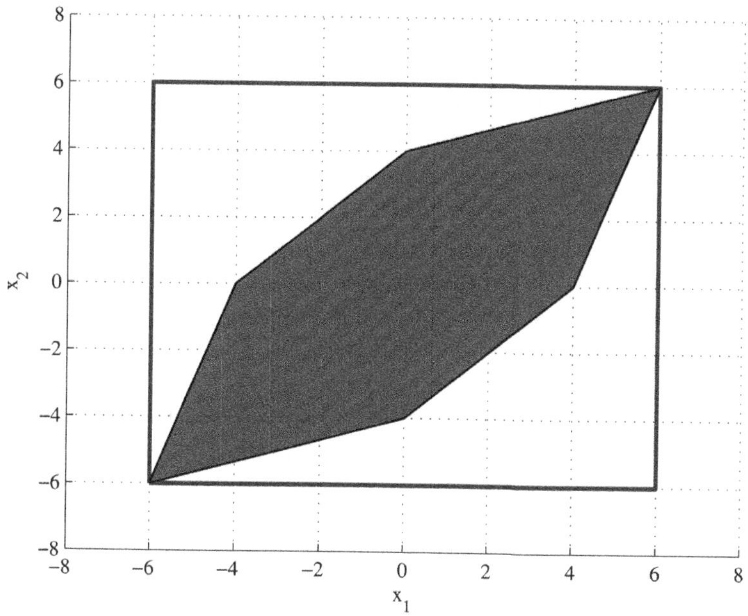

Figure 1.9. *Interval hull of a zonotope. For a color version of this figure, see www.iste.co.uk / stoica / zonotopes.zip*

In order to reduce the conservatism, an outer approximation based on the parallelotope hull is further presented.

PROPOSITION 1.2.– COMPLEXITY REDUCTION OF ZONOTOPES VIA PARALLELOTOPE HULL METHOD – Given a zonotope $\mathcal{Z} = \mathbf{p} \oplus H\mathbf{B}^m \in \mathbb{R}^n$ $(m > n)$, an over-approximation of this zonotope by a parallelotope is computed as:

$$\blacklozenge\mathcal{Z} = \Gamma \cdot \blacksquare(\Gamma^{-1}\mathcal{Z}) \qquad [1.7]$$

where $\Gamma \in \mathbb{R}^{n \times n}$ is an invertible matrix containing n columns taken from H.

PROOF [ALT 10].– This approach first transforms the coordinates of $\mathcal{Z}$ by the linear mapping Γ^{-1} (leading to

$\Gamma^{-1}\mathcal{Z} = (\Gamma^{-1}\mathbf{p}) \oplus (\Gamma^{-1}H)\mathbf{B}^m)$, where the new coordinate axes are the column vectors of Γ. In these new coordinates, the zonotope is over-approximated by a box using the interval hull. This box is transformed back to the original coordinate system, resulting in a parallelotope. The over-approximation is guaranteed by the fact that the parallelotope is over-approximated in the transformed coordinate system by the interval hull operator, such that it is also over-approximated after the transformation to the original coordinate system. □

EXAMPLE 1.10.– The same zonotope as in Example 1.9 is taken in order to compare the two approximation methods. In Figure 1.10, this zonotope $\mathcal{Z}$ is over-approximated by three different parallelotopes $\mathcal{P}_1$, $\mathcal{P}_2$, $\mathcal{P}_3$ due to the different choice of the matrix Γ ($\Gamma_1 = \begin{bmatrix} 1 & 2 \\ 3 & 2 \end{bmatrix}$, $\Gamma_2 = \begin{bmatrix} 1 & 3 \\ 3 & 1 \end{bmatrix}$, $\Gamma_3 = \begin{bmatrix} 2 & 2 \\ 3 & 1 \end{bmatrix}$). Comparing the two examples (Figures 1.9 and 1.10), the over-approximation by the parallelotope hull is less conservative than the over-approximation by interval hull, but with a higher complexity because n generators must be chosen among m generators to have the best approximation (the $\mathcal{P}_2$ parallelotope). Some criteria for selecting the suitable generators are given in [MOO 66] and [ALT 10].

PROPOSITION 1.3.– CASCADE REDUCTION OF ZONOTOPES VIA THE GENERATOR SELECTION METHOD – Given a zonotope $\mathcal{Z} = \mathbf{p} \oplus H\mathbf{B}^{m \cdot n} \in \mathbb{R}^n$ ($m \geq n$), with H a m-block matrix of $n \times n$ matrices ($H = [H_1 \dots H_m]$), let $D(l) = [H_1 \dots H_l]$ be the matrix obtained by choosing l blocks of H. Choosing the biggest l ($2 \leq l \leq m$) for which $\|D(l-1)\|_\infty > \|H_l\|_\infty$ or $l = 1$ if such an integer does not exist, this norm criterion is called the *fullness criterion* and *imposes* that the small parallelotope will be over-approximated more frequently than the big parallelotope. Then an over-approximation of $\mathcal{Z}$ is defined by:

$$\mathcal{Z} \subseteq \mathbf{p} \oplus [rs(D(l))\ H_{l+1} \dots H_m] \qquad [1.8]$$

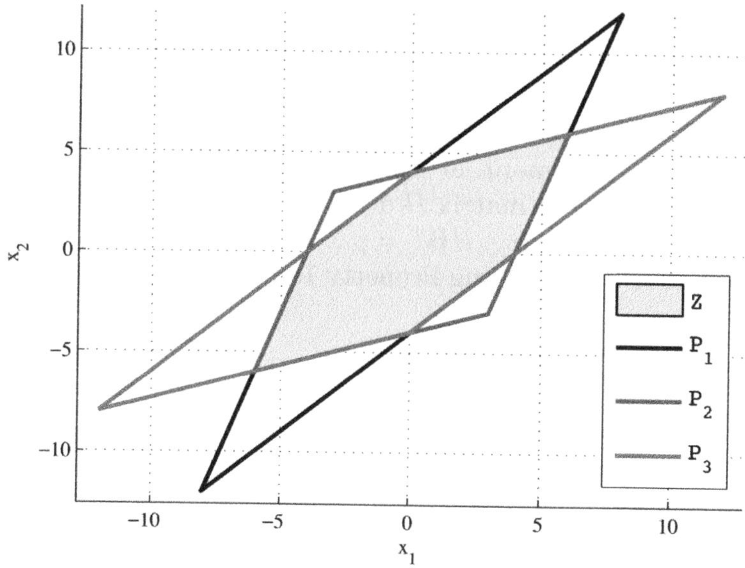

Figure 1.10. *Parallelotope hull of a zonotope. For a color version of this figure, see www.iste.co.uk / stoica / zonotopes.zip*

This proposition is based on the representation of a zonotope by the Minkowski sum of parallelotopes. The interested reader can find more details on this proposition and its performance in [KÜH 98a].

PROPOSITION 1.4.– CRITERION-BASED REDUCTION OF A ZONOTOPE – Given the zonotope $\mathcal{Z} = \mathbf{p} \oplus H\mathbf{B}^m \in \mathbb{R}^n$ and the integer s, with $n < s < m$, denote $\hat{H}$ the matrix resulting from the reordering of the columns of the matrix H by a criterion which is detailed in the following ($\hat{H} = \left[\hat{h}_1 \ldots \hat{h}_i \ldots \hat{h}_m\right]$). The zonotope is rewritten as: $\mathcal{Z} = \mathbf{p} \oplus \hat{H}_1\mathbf{B}^{s-n} \oplus \hat{H}_2\mathbf{B}^{m-s+n}$, where $\hat{H}_1$ is obtained from the first $s-n$ columns of matrix $\hat{H}$ and $\hat{H}_2$ is the remainder of $\hat{H}$. Then the initial zonotope is over-approximated by a zonotope of reduced order s as follows: $\mathcal{Z} \subseteq \mathbf{p} \oplus \hat{H}_1\mathbf{B}^{s-n} \oplus Q\mathbf{B}^n$, where $Q\mathbf{B}^n$ is the over-approximation of the zonotope $\hat{H}_2\mathbf{B}^{m-s-n}$.

This over-approximation can be a box using Proposition 1.1 or a parallelotope using Proposition 1.2.

PROOF.– [ALA 05] – Since a column of the matrix H represents a segment of the zonotope $\mathcal{Z}$, then a column permutation in the matrix H does not modify the zonotope $\mathcal{Z}$. It means that $\mathcal{Z} = \mathbf{p} \oplus H\mathbf{B}^m = p \oplus \hat{H}\mathbf{B}^m$. From the definition of matrix $\hat{H}$ and applying Property 1.4, it results in:

$$\begin{aligned} \mathcal{Z} &= \mathbf{p} \oplus \left[\hat{H}_1\ \hat{H}_2\right] \mathbf{B}^m = \\ &= \mathbf{p} \oplus \hat{H}_1\mathbf{B}^{s-n} \oplus \hat{H}_2\mathbf{B}^{m-s+n} = \\ &= \mathbf{p} \oplus \hat{H}_1\mathbf{B}^{s-n} \oplus \left[\hat{h}_{s-n+1}\ \ldots\ \hat{h}_m\right] \mathbf{B}^{m-s+n} \end{aligned} \qquad [1.9]$$

Propositions 1.1 and 1.2 show that the centered zonotope $\left[\hat{h}_{s-n+1}\ \ldots\ \hat{h}_m\right] \mathbf{B}^{m-s+n}$ can be approximated by $Q\mathbf{B}^n$, where Q is a diagonal matrix (if a box-based approximation is used) or a full matrix (if a parallelotope-based approximation is used). Therefore, the following expression is true: $\mathcal{Z} \subseteq \mathbf{p} \oplus \hat{H}_1\mathbf{B}^{s-n} \oplus Q\mathbf{B}^n = \mathbf{p} \oplus \left[\hat{H}_1\ Q\right]\mathbf{B}^s$. □

Using Proposition 1.4, the quality of the approximation depends on the value of s which limits the complexity, the criterion used to split the zonotope $\mathcal{Z}$ and the approximation method (box or parallelotope) used for the zonotope $\hat{H}_2\mathbf{B}^{m-s-n}$. A large value of s means high accuracy of the approximation and high complexity of the computation. Two methods to split the zonotope $\mathcal{Z}$ are described below.

A first approach consists of sorting the generators of the zonotope in decreasing order of the Euclidean norm

[COM 03, ALA 05], which is equivalent to arranging the segments of zonotope from the longest to the shortest. Then the longest segments which have a more important role in the shape of the zonotope are kept and the contribution of the shortest segments is over-approximated by a box or a parallelotope in order to limit the complexity.

A second approach consists of reordering the columns of matrix H in decreasing order of the term $\|h_i\|_1 - \|h_i\|_\infty$ [GIR 05]. The chosen generators (whose contribution is approximated) are close to vectors with only one non-zero component and are therefore well approximated by an interval hull.

An example is proposed in order to better illustrate the quality of these methods of complexity reduction.

EXAMPLE 1.11.– Consider a centered zonotope $\mathcal{Z} = H\mathbf{B}^8 \in \mathbb{R}^2$, with $m = 8$, $n = 2$ and $H = \begin{bmatrix} 0.9169 & 0.8936 & 0.3529 & 0.0099 & 0.2028 & 0.6038 & 0.1988 & 0.7468 \\ 0.4103 & 0.0579 & 0.8132 & 0.1389 & 0.1987 & 0.2722 & 0.0153 & 0.4451 \end{bmatrix}$.

This zonotope is approximated using Propositions 1.3 and 1.4[11]. Using the cascade reduction, the value of $l = 3$ is obtained. Figure 1.11 shows the approximation of the initial zonotope using the cascade reduction. Figure 1.12 shows the over-approximation (Z_{initial}) of the zonotope $\mathcal{Z}$ obtained using, as the criterion, the Euclidean norm with different values of s (with $s = 4$, $s = 5$). This example confirms that a bigger value of s leads to a better approximation.

11 In Proposition 1.4, the over-approximation by a box is used.

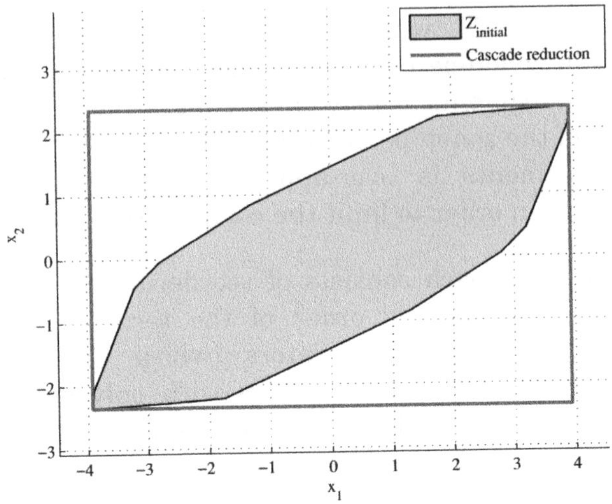

Figure 1.11. *Complexity reduction of a zonotope using the cascade reduction. For a color version of this figure, see www.iste.co.uk / stoica / zonotopes.zip*

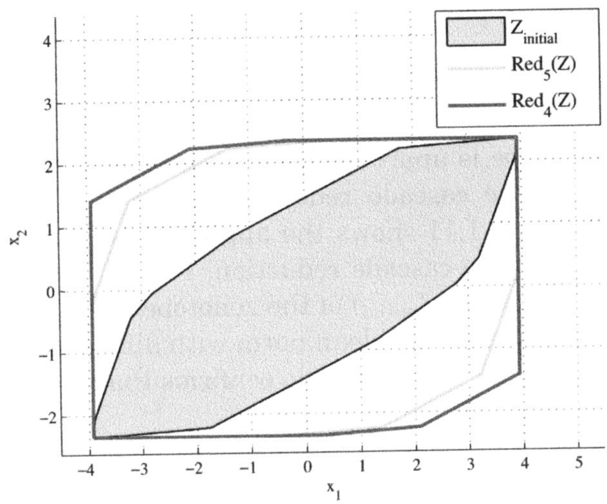

Figure 1.12. *Complexity reduction of a zonotope using the Euclidean norm-based criterion. For a color version of this figure, see www.iste.co.uk / stoica / zonotopes.zip*

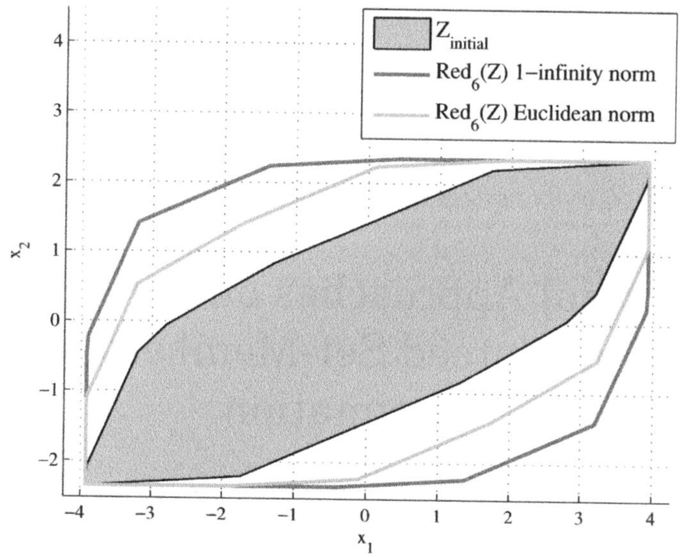

Figure 1.13. *Complexity reduction of a zonotope: comparing two criteria. For a color version of this figure, see www.iste.co.uk / stoica / zonotopes.zip*

Figure 1.13 compares the performance of the over-approximation of the same zonotope $\mathcal{Z}$ based on the two criteria of the generators' reordering and selection: the Euclidean norm and the difference between the H_1 norm and the H_∞ norm. The same value $s = 6$ is chosen for both cases. In this example, the best approximation is obtained using the Euclidean norm-based criterion. Therefore, in the further examples proposed in this book, the over-approximation based on the Euclidean norm criterion is used.

Chapter 2

Several Approaches on Zonotopic Guaranteed Set-Membership Estimation

This chapter provides an overview of the existing techniques on guaranteed set-membership estimation, more precisely related to zonotopic sets. Uncertainties acting on systems are modeled using deterministic approaches.

2.1. Context

The choice of the appropriate mathematical model, which best describes the behavior of a given plant, has received increased attention in the control system literature. Mathematical models consist of several functions describing the relationship between the system inputs, outputs and the (internal) states of the system at a given time instant. For example, the states of the mathematical model of a motor can be the current, the velocity or the position of the motor. In control systems and fault detection, knowledge about the system state is necessary for studying the system behavior and determining the control action. In practical applications, the measurement of the system states is not always available (due to the cost of the sensor or the harsh environment) or can be affected by measurement noises (sometimes introducing a significant error relative to the real values of the states). For this reason, if the system is observable, a state estimator is set up to augment or replace measurement devices in the control system. This estimator uses the knowledge of the system (mathematical model, input and output signals) to produce the estimated states. The state estimator allows us to remove the sensor, and thus reduce the cost and improve the reliability of the system with a good quality estimator. Since the 1960s, this problem has been studied by many authors, leading to different estimation techniques such as Kalman filter [KAL 60], Luenberger observer [LUE 64], set-membership estimation [WIT 68, SCH 68], functional observer [MUR 73, MOO 75] and moving-horizon estimation [GRI 90, MIC 91, MAR 06, ALA 07]. Several popular approaches discussed in the literature are listed below according to the choice of the system model.

– *Luenberger observer*: In the context of linear systems without uncertainty, a Luenberger observer can be easily implemented [LUE 64, LUE 65] due to its simplicity and efficiency.

Consider a linear discrete-time invariant system [1] in the state-space representation:

$$\begin{cases} \mathbf{x}_{k+1} = A\mathbf{x}_k + B\mathbf{u}_k \\ \mathbf{y}_k = C\mathbf{x}_k + D\mathbf{u}_k \end{cases} \qquad [2.1]$$

1 The discrete model is chosen due to its facility when systems are controlled via computer.

where $\mathbf{x}_k \in \mathbb{R}^{n_x}$ is the system state, $\mathbf{u}_k \in \mathbb{R}^{n_u}$ is the system input and $\mathbf{y}_k \in \mathbb{R}^{n_y}$ is the output measurement. The matrices A, B, C and D have appropriate dimensions $A \in \mathbb{R}^{n_x \times n_x}$, $B \in \mathbb{R}^{n_x \times n_u}$, $C \in \mathbb{R}^{n_y \times n_x}$ and $D \in \mathbb{R}^{n_y \times n_x}$. The system states can be estimated using the following observer model:

$$\begin{cases} \hat{\mathbf{x}}_{k+1} = A\hat{\mathbf{x}}_k + B\mathbf{u}_k + L(\mathbf{y}_k - \hat{\mathbf{y}}_k) \\ \hat{\mathbf{y}}_k = C\hat{\mathbf{x}}_k + D\mathbf{u}_k \end{cases} \qquad [2.2]$$

where $\hat{\mathbf{x}}_k$ and $\hat{\mathbf{y}}_k$ represent the state and output estimations at time k.

The estimation error $\mathbf{e}_k$ is then computed by:

$$\mathbf{e}_{k+1} = \mathbf{x}_{k+1} - \hat{\mathbf{x}}_{k+1} = (A - LC)\mathbf{e}_k \qquad [2.3]$$

From the estimation error equation, it is proved that the estimation error tends to 0 if the gains matrix L is chosen such that the matrix $A - LC$ is Schur stable (all its eigenvalues are inside the unit circle).

– *Kalman filter*: If uncertainties are taken into account in the mathematical model in the form of a stochastic process, the Kalman filter is presented as an effective solution [KAL 60]. A complete introduction of the Kalman filter can be found in [MAY 79]. Consider a linear discrete-time invariant system:

$$\begin{cases} \mathbf{x}_k = A\mathbf{x}_{k-1} + B\mathbf{u}_{k-1} + \omega_{k-1} \\ \mathbf{y}_k = C\mathbf{x}_k + \mathbf{v}_k \end{cases} \qquad [2.4]$$

where ω_k and $\mathbf{v}_k$ represent the process and the measurement noises, respectively. These disturbances and noises are assumed to be independent, zero mean, with normal distribution and covariance Q, R ($\omega_k \sim N(0,Q)$, $\mathbf{v}_k \sim N(0,R)$). The Kalman filter is simply an optimal stochastic recursive estimator. At each time instant, it offers the value of the estimated state $\hat{\mathbf{x}}_k$ and the error covariance matrix P_k which is a measure of the estimated accuracy of the estimated state. This is divided into two steps: prediction and update. The prediction step uses the estimated state at the previous time instant $\hat{\mathbf{x}}_{k-1}$ and the knowledge of the system (the matrices A, B and the input at the previous moment $\mathbf{u}_{k-1}$) to produce *a priori* state estimation. In the update step, this *a priori* state estimation is combined with the information from the measurement to obtain an *a posteriori* estimated state. This can be summarized in the following algorithm.

ALGORITHM 2.1.–

1) *Prediction step*:

- Compute the *a priori* state estimation $\bar{\mathbf{x}}_k = A\hat{\mathbf{x}}_{k-1} + B\mathbf{u}_{k-1}$.

- Compute the *a priori* estimation error covariance: $\bar{P}_k = AP_{k-1}A^T + Q$.

2) *Update step*:

- Compute the optimal Kalman gain: $K_k = \bar{P}_k C^T(C\bar{P}_k C^T + R)^{-1}$. This gain is determined by minimizing the mean square of the estimation error. The detail of computation can be found in [MAY 79].

- Update the state estimate $\hat{\mathbf{x}}_k = \bar{\mathbf{x}}_k + K_k(\mathbf{y}_k - C\bar{\mathbf{x}}_k)$.

- Update the estimate error covariance $P_k = (I - K_k C)\bar{P}_k$.

For the interested reader, a toolbox dedicated to this problem is developed[2]. In the parametric uncertainty context, it is well known that the performance of the proposed Kalman filter can be degraded [BRO 97]. To ensure the convergence behavior in the presence of a modeling error, the robust Kalman filter is proposed, which guarantees a

2 See http://www.cs.ubc.ca/~murphyk/Software/Kalman/kalman.html.

bound of the performance of the Kalman filter [XIE 94, STE 95, OHR 95, FOO 08].

– *Set-membership estimation*: when a system is modeled by the deterministic approach (uncertainties are bounded by some convex sets), the set-membership estimation is a suitable solution for the state estimation problem. This technique has been developed in the past 35 years [WIT 68, SCH 68, BER 71]. The estimator computes, at each sample time, a set containing all the possible system states that are consistent with the perturbations, the uncertainties and the measurement noises. Based on the prediction and correction steps, the procedure of this technique is similar to the Kalman filter. While the Kalman filter deals with the average case, the set-membership estimation considers the worst case. For this reason, this approach is also called the worst-case estimation. The problem of set-membership estimation is that the complexity of this set increases over time. To overcome this problem, the geometry of these sets has to be fixed *a priori*: for example, polytopes (boxes and parallelotopes) [WAL 89, VIC 96, CHI 96, BEM 00, GAR 01, JAU 01, KIE 02, PUI 03, RAÏ 04, MES 10], ellipsoids [SCH 68, WIT 68, BER 71, KUR 96, GHA 99, DUR 01, POL 04, BEC 08, BEC 11, BEC 12] and zonotopes [PUI 01, COM 03, ALA 05, ALA 08, LE 11a]. Polytopes which were presented in the previous chapter offer a good quality of approximation. In the linear context, polytopes can be used for an exact representation of the variation domains of the system state. However, efficient results may be obtained only for a reasonable number of vertices of the polytopes [WAL 89]. Because of the low complexity, ellipsoids have been used by many authors but their limited flexibility can lead to a conservative result of estimation. As presented in the previous chapter, zonotopes which are used in many automatic control branches, such as reachability analysis [ALT 07], collision detection [GUI 05], identification [BRA 06], state estimation [PUI 01, COM 03, ALA 05,

ALA 08], fault detection [GUE 08, ING 08, STO 11] and fault diagnosis [COM 08], offer a good compromise between complexity and tflexibility. Moreover, the author of [KÜH 98a] shows that, by using zonotopes, the wrapping effect is reduced leading to a more precise result of the estimation. As the deterministic approach is chosen to describe the modeling of the system, the set-membership estimation is considered as a suitable solution. In the following section, zonotopes are chosen to represent the set of all the possible system states in the context of set-membership estimation due to their advantages in comparison with other geometrical forms.

2.2. Problem formulation

To simplify the chapter, the following linear discrete-time autonomous system is considered (this system can be easily generalized to controlled system):

$$\begin{cases} \mathbf{x}_{k+1} = A\mathbf{x}_k + \omega_k \\ \mathbf{y}_k = C\mathbf{x}_k + \mathbf{v}_k \end{cases} \qquad [2.5]$$

where $\mathbf{x}_k \in \mathbb{R}^{n_x}$ is the state of the system, $\mathbf{y}_k \in \mathbb{R}^{n_y}$ is the measured output at sample time k, the matrices A and C have appropriate dimensions $A \in \mathbb{R}^{n_x \times n_x}$ and $C \in \mathbb{R}^{n_y \times n_x}$ and the couple (C,A) is detectable. The vector $\omega_k \in \mathbb{R}^{n_x}$ represents the state perturbation vector and $\mathbf{v}_k \in \mathbb{R}^{n_y}$ is the measurement perturbation (noise, offset, etc.). It is assumed that the uncertainties and the initial state are bounded by zonotopes: $\omega_k \in \mathcal{W}, \mathbf{v}_k \in \mathcal{V}$ and the initial state belongs to a zonotope $\mathbf{x}_0 \in \mathcal{X}_0$ which can be large due to the lack of knowledge of the system. The two zonotopes $\mathcal{W}$ and $\mathcal{V}$ are assumed to be centered at the origin; if this assumption is not satisfied, an appropriate change of coordinates can be used to bring the center of the zonotopes to the origin.

Consider the mathematical model [2.5] and these assumptions, the set-membership estimation technique leads to the computation of, at each sample time k, a domain of all the possible values of the unknown state $\mathbf{x}_k$. Similar to the Kalman filter, the set-membership estimation algorithm is based on three steps: prediction, measurement and correction[3]. The guaranteed state estimation is obtained in the correction step which is the combination of the state information from the prediction and the information from the measurement. Before detailing this algorithm, some useful notations will be defined.

DEFINITION 2.1.– *Given the system* [2.5] *and a measured output* $\boldsymbol{y}_k$, *the* measurement consistent state set *at time instant* k *(the state set which is consistent with the measured output* $\boldsymbol{y}_k$*) is defined as* $\mathcal{X}_{\boldsymbol{y}_k} = \{\boldsymbol{x} \in \mathbb{R}^n : (\boldsymbol{y}_k - C\boldsymbol{x}) \in \mathcal{V}\}$.

DEFINITION 2.2.– *Consider the system* [2.5]. *The* exact uncertain state set $\mathcal{X}_k = (A\mathcal{X}_{k-1} \oplus \mathcal{W}) \cap \mathcal{X}_{\boldsymbol{y}_k}$, $k \geq 1$ *is equal to the set of states that are consistent with the measured output and the initial state set* $\mathcal{X}_0$.

Thus, the exact uncertain state set $\mathcal{X}_k$ contains all the possible values of the system state consistent with the measurement. In practice, the computation of the exact uncertain state set is difficult. Even if $\mathcal{X}_{k-1}$ is assumed to have a particular geometrical form (i.e. zonotope and ellipsoid), it is not certain that at time instant k the exact uncertain state set $\mathcal{X}_k$ has the same form. For this reason, in practice, this set is approximated by an outer bound[4]. The following assumptions are considered at time instant k:

---

3 In the Kalman filter, the measurement step is included in the correction step, in the set-membership algorithm it is clear-cut for a better understanding.

4 Zonotopic outer approximations will be used in this book.

– An outer bound of the exact uncertain state set, denoted by $\hat{\mathcal{X}}_{k-1}$, is available.

– An output measurement $\mathbf{y}_k$ is obtained.

Under these assumptions, a zonotopic outer bound $\hat{\mathcal{X}}_k$ of the exact uncertain state set $\mathcal{X}_k$ can be estimated using the following algorithm.

ALGORITHM 2.2.–

1) *Prediction step*: given the system [2.5], compute a zonotope $\bar{\mathcal{X}}_k = A\hat{\mathcal{X}}_{k-1} \oplus \mathcal{W}$ (denoted *predicted state set*) that offers a bound for the uncertain trajectory of the system.

2) *Measurement step*: compute the *measurement consistent state set* $\mathcal{X}_{\mathbf{y}_k}$ using the measurement $\mathbf{y}_k$.

3) *Correction step*: compute an outer approximation $\hat{\mathcal{X}}_k$ (denoted *guaranteed state estimation set*) of the intersection between $\mathcal{X}_{\mathbf{y}_k}$ and $\bar{\mathcal{X}}_k$.

The general case of the proposed algorithm is illustrated in Figure 2.1. At time instant k, from the known guaranteed state estimation $\hat{\mathcal{X}}_{k-1}$, the predicted state set $\bar{\mathcal{X}}_k$ is determined. This set is intersected with the measurement consistent state set $\mathcal{X}_{\mathbf{y}_k}$ which is induced by the output measurement $\mathbf{y}_k$. Thus, at time instant k, the guaranteed state estimation $\hat{\mathcal{X}}_k$, is the outer approximation of this intersection.

To obtain a zonotope bounding the uncertain trajectory of the system in step 1 of Algorithm 2.2, Properties 1.2 and 1.3 are used. The predicted state set computation using zonotopes relies on a simple matrix computation and is not subject to any approximation. However, this computation increases the order of the zonotope at each step. In order to control the domain complexity, a reduction step is implemented to bound a high-order zonotope by a low-order zonotope using the complexity reduction method presented in Chapter 1.

In the literature, there are several authors interested in this zonotopic set-membership estimation problem [COM 03, COM 04, COM 05, ALA 05, ALA 08]. In the linear context, the first step of Algorithm 2.2 is similar in these methods. Based on different methods for realizing the correction step (the intersection between the predicted state set and the measurement consistent state set), two different approaches presented in the literature will be further detailed: the Singular Value Decomposition (SVD)-based method [COM 03] and the optimization-based method [ALA 05].

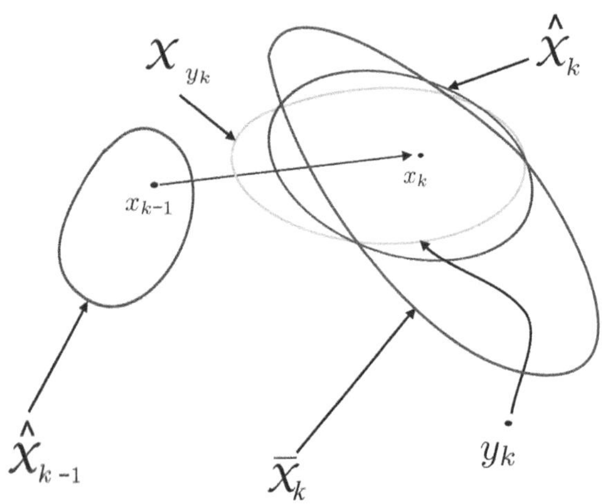

Figure 2.1. *Set-membership estimation algorithm. For a color version of this figure, see www.iste.co.uk / stoica / zonotopes.zip*

2.2.1. *Singular Value Decomposition-based method*

This method was first proposed in [COM 03]. The zonotope bounding the disturbance is assumed to be a centered zonotope represented by $\mathcal{W} = F\mathbf{B}^{n_x}$. In this method, the measurement noise is assumed to belong to a centered parallelotope $\mathcal{V}$, which can be described by $\mathcal{V} = \Sigma\mathbf{B}^{n_y}$, with $\Sigma \in \mathbb{R}^{n_y \times n_y}$ an invertible matrix. With these notations, at each sample time k, there exists a vector $\mathbf{b}_k \in \mathbf{B}^{n_y}$ such that the measurement noise at time instant k is computed by $\mathbf{v}_k = \Sigma \mathbf{b}_k$.

Let us suppose that at time instant k the zonotopic guaranteed state estimation at $k-1$ is available: $\hat{\mathcal{X}}_{k-1} = \hat{\mathbf{p}}_{k-1} \oplus \hat{H}_{k-1}\mathbf{B}^r$, $r \in \mathbb{R}^+$. Using the mathematical model, the predicted state set $\bar{\mathcal{X}}_k$ in the first step of Algorithm 2.2, which contains the real state $\mathbf{x}_k$, can be computed by:

$$\begin{aligned} \bar{\mathcal{X}}_k &= A\hat{\mathcal{X}}_{k-1} \oplus \mathcal{W} \\ &= A(\hat{\mathbf{p}}_{k-1} \oplus \hat{H}_{k-1}\mathbf{B}^r) \oplus F\mathbf{B}^{n_x} \\ &= A\hat{\mathbf{p}}_{k-1} \oplus A\hat{H}_{k-1}\mathbf{B}^r \oplus F\mathbf{B}^{n_x} \\ &= A\hat{\mathbf{p}}_{k-1} \oplus \left[A\hat{H}_{k-1}\ F\right]\mathbf{B}^{r+n_x} \\ &= \bar{\mathbf{p}}_k \oplus \bar{H}_k\mathbf{B}^{r+n_x} \end{aligned} \quad [2.6]$$

with $\bar{\mathbf{p}}_k = A\hat{\mathbf{p}}_{k-1}$ and $\bar{H}_k = \left[A\hat{H}_{k-1}\ F\right]$.

From this point onwards, in this chapter, the notation $\hat{*}$ will be used to specify the estimation of $*$ and $\bar{*}$ will be used for the prediction of $*$.

The guaranteed state estimation is obtained from the intersection between the measurement consistent state set $\mathcal{X}_{\mathbf{y}_k}$ and the predicted state set $\bar{\mathcal{X}}_k$. The method proposed in [COM 03] considers a slightly different approach in which the measurement consistent state set $\mathcal{X}_{\mathbf{y}_k}$ is not explicitly computed.

This method consists of decomposing the extended space $\mathbb{R}^{r+n_x}$ (called the *abstract space* in [COM 03]) of $\bar{\mathcal{X}}_k$ into two complementary subspaces using the SVD: one subspace is influenced by the measurement, while the other subspace is not influenced. Uniquely, the outer approximation in the subspace containing the information coming from the measurement is considered.

A prediction of the measurement at time instant k can be obtained from the center of $\hat{\mathcal{X}}_{k-1}$:

$$\bar{\mathbf{y}}_k = C\bar{\mathbf{p}}_k \qquad [2.7]$$

The real measurement is given by [2.5]:

$$\mathbf{y}_k = C\mathbf{x}_k + \Sigma\mathbf{b}_k \qquad [2.8]$$

The difference between $\mathbf{y}_k$ and $\bar{\mathbf{y}}_k$ reflects the supplementary information coming from the measurement:

$$\mathbf{y}_k - \bar{\mathbf{y}}_k = C(\mathbf{x}_k - \bar{\mathbf{p}}_k) + \Sigma\mathbf{b}_k \qquad [2.9]$$

As Σ is an invertible matrix, by multiplying [2.9] with Σ^{-1} this is equivalent to:

$$\Sigma^{-1}(\mathbf{y}_k - \bar{\mathbf{y}}_k) = \Sigma^{-1}C(\mathbf{x}_k - \bar{\mathbf{p}}_k) + \mathbf{b}_k \qquad [2.10]$$

Denoted by $\eta = \Sigma^{-1}(\mathbf{y}_k - \bar{\mathbf{y}}_k)$ and $M = \Sigma^{-1}C$. Because the real state $\mathbf{x}_k$ belongs to the predicted state set $\bar{\mathcal{X}}_k$ formulated by [2.6], a vector $\mathbf{s} \in \mathbf{B}^{r+n_x}$ exists such that $\mathbf{x}_k = \bar{\mathbf{p}}_k + \bar{H}_k\mathbf{s}$. Thus, the following equation is verified:

$$\begin{aligned}\eta - \mathbf{b}_k &= M(\mathbf{x}_k - \bar{\mathbf{p}}_k) \\ &= M(\bar{\mathbf{p}}_k + \bar{H}_k\mathbf{s} - \bar{\mathbf{p}}_k) \\ &= M\bar{H}_k\mathbf{s}\end{aligned} \qquad [2.11]$$

The correction step focuses on the outer approximation of the intersection between $\mathbf{B}^{r+n_x}$ and the domain of possible values of $\mathbf{s}$ resulting from the measurement equation [2.8].

The procedure proposed in [COM 03] for computing the zonotopic guaranteed state estimation is based on the SVD of the matrix that resulted from the product $M\bar{H}_k \in \mathbb{R}^{n_y \times (r+n_x)}$.

Find the SVD of $M\bar{H}_k$:

$$M\bar{H}_k = USV^T = \begin{bmatrix} U_1 & U_0 \end{bmatrix} \begin{bmatrix} S_1 & 0 \\ 0 & 0 \end{bmatrix} \begin{bmatrix} V_1^T \\ V_0^T \end{bmatrix} \quad [2.12]$$

with $U^TU = I$ and $V^TV = I$ and S_1 a diagonal matrix with non-zero elements which are the singular values of $M\bar{H}_k$.

The initial measurement space is generated by U_0 and U_1 and the abstract space is generated by V_0 and V_1 [COM 03]. The subspace generated by V_0 is the Kernel of $M\bar{H}_k$ and is not influenced by the output measurement. The subspace influenced by the output measurement is generated by V_1.

In fact, V_0 and V_1 are the new bases and, thus, in the new base, the vector $\mathbf{s}$ can be decomposed as:

$$\mathbf{s} = V_0\boldsymbol{\delta}_0 + V_1\boldsymbol{\delta}_1 \quad [2.13]$$

and $\boldsymbol{\delta}_0 = V_0^T\mathbf{s}$, $\boldsymbol{\delta}_1 = V_1^T\mathbf{s}$ with $\boldsymbol{\delta}_0$, $\boldsymbol{\delta}_1$ the coordinate of $\mathbf{s}$ in the new base. With these new notations, equation [2.11] is equivalent to:

$$\boldsymbol{\eta} - \mathbf{b}_k = M\bar{H}_k(V_0\boldsymbol{\delta}_0 + V_1\boldsymbol{\delta}_1) \quad [2.14]$$

Replacing $M\bar{H}_k$ with its SVD leads to:

$$\boldsymbol{\eta} - \mathbf{b}_k = U_1S_1V_1^T(V_0\boldsymbol{\delta}_0 + V_1\boldsymbol{\delta}_1) \quad [2.15]$$

From the expressions $U^TU = I$, $V^TV = I$, it leads to $U_1^TU_1 = I$, $V_0^TV_0 = I$, $V_1^TV_1 = I$, $V_1^TV_0 = 0$, $V_0^TV_1 = 0$. Thus, equation [2.15] is equivalent to:

$$\boldsymbol{\delta}_1 = S_1^{-1}U_1^T\boldsymbol{\eta} - S_1^{-1}U_1^T\mathbf{b}_k \quad [2.16]$$

Because $\mathbf{b}_k$ is an interval vector, from the definition of zonotope, a sufficient condition for equation [2.16] or [2.11] is the following:

$$\boldsymbol{\delta}_1 \in \mathcal{Z}_1 = \mathcal{Z}(S_1^{-1}U_1^T\boldsymbol{\eta}; S_1^{-1}U_1^T) \quad [2.17]$$

From equation [2.13], the images of $\mathbf{s}$ in the subspace generated by V_1 and the subspace generated by V_0 are shown by:

$$\boldsymbol{\delta}_1 \in \mathcal{Z}(0; V_1^T), \boldsymbol{\delta}_0 \in \mathcal{Z}(0; V_0^T) \quad [2.18]$$

Combining the results of [2.17] and [2.18] leads to:

$$\boldsymbol{\delta}_1 \in \mathcal{Z}_1 \cap \mathcal{Z}(0; V_1^T), \boldsymbol{\delta}_0 \in \mathcal{Z}(0; V_0^T) \quad [2.19]$$

The next step consists of finding the outer approximation of the intersection between $\mathcal{Z}_1$ and $\mathcal{Z}(0; V_1^T)$. This can be computed as the intersection of the interval hulls of the two zonotopes $\mathcal{Z}_1$ and $\mathcal{Z}(0; V_1^T)$ (i.e. the intersection of two boxes in this case) which is an easy piece of work:

$$\mathcal{Z}(\mathbf{p}_{temp}; H_{temp}) = \mathcal{Z}(S_1^{-1}U_1^T\boldsymbol{\eta}; rs(S_1^{-1}U_1^T)) \cap \mathcal{Z}(0; rs(V_1^T)) \quad [2.20]$$

Returning to the abstract space [2.13], the approximation of $\mathbf{s}$ can be made by combining the information from [2.19] and [2.20]:

$$\mathbf{s} \in \mathcal{Z}(V_1\mathbf{p}_{temp}; \left[V_1 H_{temp} \; V_0 V_0^T\right]) \quad [2.21]$$

Thus, the guaranteed state estimation at time instant k is determined by:

$$\hat{\mathcal{X}}_k = \mathcal{Z}(\bar{\mathbf{p}} + \bar{H}_k V_1 \mathbf{p}_{temp}; \left[\bar{H}_k V_1 H_{temp} \; \bar{H}_k V_0 V_0^T\right]) \quad [2.22]$$

Detailed explanations about this method can be found in [COM 03] and [COM 05]. An improved version of this approach which consists of replacing the intersection of two boxes [2.20] by a zonotopic outer approximation of the intersection between two zonotopes is presented in [COM 04]. The SVD method allows us to rapidly obtain a guaranteed state estimation but it cannot guarantee that the size of this guaranteed state estimation is optimized. For this reason, another method based on the solution of an optimization problem is presented in the following section.

2.2.2. *Optimization-based methods*

This method presented in [ALA 05] is based on the zonotopic approximation of the intersection between a zonotope and a strip [5]. The algorithm presented in [ALA 05] is developed for single output systems ($n_y = 1$ in equation [2.5]). The disturbances are bounded by a centered zonotope $\mathcal{W} = F\mathbf{B}^{n_x}$. The measurement noise is supposed to belong to a centered interval $\mathcal{V} = \sigma\mathbf{B}^1$, $\sigma \in \mathbb{R}^+$. With this assumption, from the mathematical model [2.5], the output measurement can be written as $y_k = \mathbf{c}^T\mathbf{x}_k + v_k$, with $v_k \in \mathcal{V}$ and $\mathbf{c} \in \mathbb{R}^{n_x}$.

Suppose that the guaranteed state estimation at time instant $k - 1$ is known at time instant k (i.e. $\hat{\mathcal{X}}_{k-1} = \hat{\mathbf{p}}_{k-1} \oplus \hat{H}_{k-1}\mathbf{B}^r$). Similar to equation [2.6], the predicted state set is determined as:

$$\bar{\mathcal{X}}_k = A\hat{\mathbf{p}}_{k-1} \oplus \left[A\hat{H}_{k-1} \; F\right] \mathbf{B}^{r+n_x} = \bar{\mathbf{p}}_k \oplus \bar{H}_k\mathbf{B}^{r+n_x} \qquad [2.23]$$

In this case, the measurement consistent state set is computed as a strip:

$$\mathcal{X}_{\mathbf{y}_k} = \{|\mathbf{c}^T\mathbf{x} - y_k| \leq \sigma\} \qquad [2.24]$$

---

5 A *strip* is defined as the set $\{\mathbf{x} \in \mathbb{R}^n : |\mathbf{c}^T\mathbf{x} - y| \leq \sigma\}$, with $\mathbf{c} \in \mathbb{R}^n$ and $y, \sigma \in \mathbb{R}$.

The guaranteed state estimation $\hat{\mathcal{X}}_k$ can be obtained by intersecting the predicted state set $\bar{\mathcal{X}}_k$ with the measurement consistent state set $\mathcal{X}_{\mathbf{y}_k}$. As $\bar{\mathcal{X}}_k$ is a zonotope and $\mathcal{X}_{\mathbf{y}_k}$ is a strip, it is convenient to obtain a zonotopic outer bound of the intersection of a zonotope and a strip. The following proposition provides a family of zonotopes (parameterized by a vector $\boldsymbol{\lambda}$) that contains the intersection of a zonotope and a strip.

PROPOSITION 2.1.– [ALA 05] – Given the zonotope $\mathcal{Z} = \mathbf{p} \oplus H\mathbf{B}^r \subset \mathbb{R}^n$, the strip $\mathcal{S} = \{\mathbf{x} \in \mathbb{R}^n : |\mathbf{c}^T\mathbf{x} - d| \leq \sigma\}$ and the vector $\boldsymbol{\lambda} \in \mathbb{R}^n$ define:

– a vector $\hat{\mathbf{p}}(\boldsymbol{\lambda}) = \mathbf{p} + \boldsymbol{\lambda}(d - \mathbf{c}^T\mathbf{p}) \in \mathbb{R}^n$;

– a matrix $\hat{H}(\boldsymbol{\lambda}) = [(I - \boldsymbol{\lambda}\mathbf{c}^T)H \ \sigma\boldsymbol{\lambda}] \in \mathbb{R}^{n\times(r+1)}$;

then the following expression holds $\mathcal{Z} \cap \mathcal{S} \subseteq \hat{\mathcal{Z}}(\boldsymbol{\lambda}) = \hat{\mathbf{p}}(\boldsymbol{\lambda}) \oplus \hat{H}(\boldsymbol{\lambda})\mathbf{B}^{r+1}$.

PROOF.– Given an element $\mathbf{x} \in \mathcal{Z} \cap \mathcal{S}$, on the one hand this means that $\mathbf{x} \in \mathcal{Z} = \mathbf{p} \oplus H\mathbf{B}^r$. Using the definition of a m-zonotope implies that there exists a vector $\mathbf{s} \in \mathbf{B}^r$ such that:

$$\mathbf{x} = \mathbf{p} + H\mathbf{s} \tag{2.25}$$

Adding and subtracting $\boldsymbol{\lambda}\mathbf{c}^T H\mathbf{s}$ to the previous equality lead to the following expression:

$$\mathbf{x} = \mathbf{p} + \boldsymbol{\lambda}\mathbf{c}^T H\mathbf{s} + (I - \boldsymbol{\lambda}\mathbf{c}^T)H\mathbf{s} \tag{2.26}$$

On the other hand, $\mathbf{x} \in \mathcal{Z} \cap \mathcal{S}$ leads to $\mathbf{x} \in \mathcal{S} = \{\mathbf{x} \in \mathbb{R}^n : |\mathbf{c}^T\mathbf{x} - d| \leq \sigma\}$. Thus, there exists a value $\alpha \in [-1; 1]$ such that $\mathbf{c}^T\mathbf{x} - d = \sigma\alpha$. Taking into account the form of the vector $\mathbf{x}$ given by [2.25] leads to $\mathbf{c}^T(\mathbf{p} + H\mathbf{s}) - d = \sigma\alpha$, which is equivalent to

$\mathbf{c}^T H\mathbf{s} = d - \mathbf{c}^T\mathbf{p} + \sigma\alpha$. Substituting $\mathbf{c}^T H\mathbf{s}$ in equation [2.26], the following expression is obtained:

$$\begin{aligned}\mathbf{x} &= \mathbf{p} + \boldsymbol{\lambda}(d - \mathbf{c}^T\mathbf{p} + \sigma\alpha) + (I - \boldsymbol{\lambda}\mathbf{c}^T)H\mathbf{s} \\ &= \mathbf{p} + \boldsymbol{\lambda}(d - \mathbf{c}^T\mathbf{p}) + \boldsymbol{\lambda}\sigma\alpha + (I - \boldsymbol{\lambda}\mathbf{c}^T)H\mathbf{s}\end{aligned} \qquad [2.27]$$

After simple computations and using the notation defined in Proposition 2.1, the following form is obtained:

$$\begin{aligned}\mathbf{x} &= \hat{\mathbf{p}}(\boldsymbol{\lambda}) + \left[(I - \boldsymbol{\lambda}\mathbf{c}^T)H \; \sigma\boldsymbol{\lambda}\right]\begin{bmatrix}\mathbf{s}\\ \alpha\end{bmatrix} = \\ &= \hat{\mathbf{p}}(\boldsymbol{\lambda}) \oplus \hat{H}(\boldsymbol{\lambda})\begin{bmatrix}\mathbf{s}\\ \alpha\end{bmatrix}\end{aligned} \qquad [2.28]$$

and the following inclusion holds:

$$\mathbf{x} = \hat{\mathbf{p}}(\boldsymbol{\lambda}) \oplus \hat{H}(\boldsymbol{\lambda})\begin{bmatrix}\mathbf{s}\\ \alpha\end{bmatrix} \in \hat{\mathbf{p}}(\boldsymbol{\lambda}) \oplus \hat{H}(\boldsymbol{\lambda})\mathbf{B}^{r+1} = \hat{\mathcal{Z}}(\boldsymbol{\lambda}) \qquad [2.29]$$

This method can be extended to the case of Multi-Output systems. Thus, considering each measurement as a strip in the state-space, the guaranteed state estimation is obtained by repeating the intersection with each strip of measurement.

Using Proposition 2.1, the guaranteed state estimation at time instant k, is the outer approximation of $\bar{\mathcal{X}}_k \cap \mathcal{X}_{\mathbf{y}_k}$, formulated by:

$$\hat{\mathcal{X}}_k(\boldsymbol{\lambda}) = \hat{\mathbf{p}}_k(\boldsymbol{\lambda}) \oplus \hat{H}_k(\boldsymbol{\lambda})\mathbf{B}^{r+n_x+1} \qquad [2.30]$$

with $\hat{\mathbf{p}}_k(\boldsymbol{\lambda}) = A\hat{\mathbf{p}}_{k-1} + \boldsymbol{\lambda}(y_k - \mathbf{c}^T A\hat{\mathbf{p}}_{k-1})$ and $\hat{H}_k(\boldsymbol{\lambda}) = \left[(I - \boldsymbol{\lambda}\mathbf{c}^T)A\hat{H}_{k-1} \; (I - \boldsymbol{\lambda}\mathbf{c}^T)F \; \sigma\boldsymbol{\lambda}\right]$.

This equation describes a family of zonotopes $\hat{\mathcal{X}}_k(\boldsymbol{\lambda})$ parameterized by the vector $\boldsymbol{\lambda}$, which bounds the intersection between $\bar{\mathcal{X}}_k$ and $\mathcal{X}_{\mathbf{y}_k}$. The vector $\boldsymbol{\lambda}$ is then determined in order to optimize the size of this zonotope $\hat{\mathcal{X}}_k(\boldsymbol{\lambda})$. In the following, three size-based criteria to compute $\boldsymbol{\lambda}$ will be presented: the minimization of the segments of the zonotope $\hat{\mathcal{X}}_k(\boldsymbol{\lambda})$, the minimization of the volume of the zonotope $\hat{\mathcal{X}}_k(\boldsymbol{\lambda})$ and the minimization of the P-radius of a zonotope.

2.2.2.1. *Minimizing the segments of the zonotope*

In this approach proposed in [ALA 05], the vector $\boldsymbol{\lambda}$ is computed such that the Sum of Squares (SOS) of the generators of the zonotope $\hat{\mathcal{X}}_k(\boldsymbol{\lambda})$ is minimized (i.e. the SOS of the segments of the zonotope). This is equivalent to minimizing the Frobenius norm of the matrix $\hat{H}_k(\boldsymbol{\lambda})$. It is convenient to decompose this matrix in the following form [ALA 05]:

$$\hat{H}_k(\boldsymbol{\lambda}) = M + \boldsymbol{\lambda}\mathbf{a}^T \qquad [2.31]$$

with $M = \begin{bmatrix}\bar{H}_k \ 0\end{bmatrix}$, $\mathbf{a}^T = \begin{bmatrix}-\mathbf{c}^T\bar{H}_k \ \sigma\end{bmatrix}$.

The Frobenius norm of $\hat{H}_k(\boldsymbol{\lambda})$ is computed by:

$$\begin{aligned}\|\hat{H}_k(\boldsymbol{\lambda})\|_F^2 &= \|M + \boldsymbol{\lambda}\mathbf{a}^T\|_F^2 \\ &= tr((M^T + \mathbf{a}\boldsymbol{\lambda}^T)(M + \boldsymbol{\lambda}\mathbf{a}^T)) \\ &= tr(M^TM) + tr(\mathbf{a}\boldsymbol{\lambda}^TM) + tr(M^T\boldsymbol{\lambda}\mathbf{a}^T) + tr(\mathbf{a}\boldsymbol{\lambda}^T\boldsymbol{\lambda}\mathbf{a}^T) \\ &= 2\boldsymbol{\lambda}^TM\mathbf{a} + \mathbf{a}^T\mathbf{a}\boldsymbol{\lambda}^T\boldsymbol{\lambda} + tr(M^TM)\end{aligned} \qquad [2.32]$$

The minimum of $\|\hat{H}_k(\boldsymbol{\lambda})\|_F^2$ is obtained when $\frac{d\|\hat{H}_k(\boldsymbol{\lambda})\|_F^2}{d\boldsymbol{\lambda}} = 0$. This means that:

$$\frac{d(2\boldsymbol{\lambda}^TM\mathbf{a} + \mathbf{a}^T\mathbf{a}\boldsymbol{\lambda}^T\boldsymbol{\lambda} + tr(M^TM))}{d\boldsymbol{\lambda}} = 0 \qquad [2.33]$$

or:

$$2M\mathbf{a} + 2\mathbf{a}^T\mathbf{a}\boldsymbol{\lambda}^* = 0 \qquad [2.34]$$

The optimal value of vector $\boldsymbol{\lambda}$ is then computed by:

$$\begin{aligned}\boldsymbol{\lambda}^* &= \frac{-M\mathbf{a}}{\mathbf{a}^T\mathbf{a}} \\ &= \frac{\bar{H}_k\bar{H}_k^T\mathbf{c}}{\mathbf{c}^T\bar{H}_k\bar{H}_k^T\mathbf{c}+\sigma^2}\end{aligned} \qquad [2.35]$$

This method allows us a fast computation of the vector $\boldsymbol{\lambda}$ which can be used in fast real-time systems; however, the result of approximation is sometimes conservative as illustrated in [ALA 05].

2.2.2.2. *Minimizing the volume of the zonotope*

In order to improve the performance of the guaranteed state estimation, another criterion is proposed. The vector $\boldsymbol{\lambda}$ is determined such that the volume of the zonotope $\hat{\mathcal{X}}_k(\boldsymbol{\lambda})$ is minimized. The volume of a zonotope $\mathcal{Z} = \mathbf{p} \oplus H\mathbf{B}^m \in \mathbb{R}^n$, with $m \geq n$ is given by the following formula [SHE 74, MON 89]:

$$Vol(\mathcal{Z}) = 2^n \sum_{i=1}^{\binom{n}{m}} |det\left[H_{s_1(i)}\ H_{s_2(i)}\ \cdots\ H_{s_n(i)}\right]| \qquad [2.36]$$

with $\binom{n}{m}$ the number of all the different ways of choosing n elements between m elements, H_i the ith column of H and $s_j(i)$, with $j = 1, \ldots, n$ and $i = 1, \ldots, \binom{n}{m}$, denotes each one of different ways of choosing n elements from a set of m elements. These integers satisfy $1 \leq s_1(i) < s_2(i) < \ldots < s_n(i) \leq m$.

Using this formula to compute the volume of the zonotope $\hat{\mathcal{X}}_k(\boldsymbol{\lambda}) = \hat{\mathbf{p}}_k(\boldsymbol{\lambda}) \oplus \hat{H}_k(\boldsymbol{\lambda})\mathbf{B}^{r+n_x+1}$ leads to:

$$Vol(\hat{\mathcal{X}}_k(\boldsymbol{\lambda})) = 2^{n_x} \sum_{i=1}^{\binom{n_x}{r+n_x}} |1 - \mathbf{c}^T\boldsymbol{\lambda}| \cdot |det(D_i)| + 2^{n_x} \sum_{i=1}^{\binom{n_x-1}{r+n_x}} \sigma|det\left[E_i \; \mathbf{q}_i\right]| \cdot |\mathbf{q}_i^T\sigma| \qquad [2.37]$$

where D_i is each of the different matrices obtained by choosing n_x columns of matrix $\bar{H}_k$, E_i is each of the different matrices obtained by choosing $n_x - 1$ columns of matrix $\bar{H}_k$ and $\mathbf{q}_i$ is orthonormal to $\mathrm{Im}(E_i)$ with $\mathbf{q}_i^T\mathbf{q}_i = 1$ and $\mathbf{q}_i^T E_i = 0$. The proof of this formula is presented in [ALA 05].

The vector $\boldsymbol{\lambda}$ is chosen to minimize the volume of the zonotope computed by [2.37]. As the volume of $\hat{\mathcal{X}}_k(\boldsymbol{\lambda})$ is a convex function of $\boldsymbol{\lambda}$, the optimal vector $\boldsymbol{\lambda}^*$ which minimizes the volume of $\hat{\mathcal{X}}_k(\boldsymbol{\lambda})$ can be found by solving a convex optimization problem. This volume-based criterion gives an improved result of the approximation in comparison to the segment-based criterion. But the complexity of equation [2.37] leads to a considerable increase of the computation time. Moreover, minimizing the volume of the zonotope can lead to a very narrow zonotope (i.e. the uncertainty in some directions can remain extremely large, even when the volume of the zonotope tends to zero). For these reasons, in the following section, a P-radius-based approach will be proposed, allowing us to obtain a good result with a low complexity.

2.2.2.3. *Minimizing the P-radius of the zonotope*

The presented approaches have their advantages and their drawbacks. The SVD method and the minimization of the volume of the zonotope lead to good results with a complex online computation. The minimization of the segments of the zonotope offers a fast online computation time with a conservative result. For these reasons, a major challenge is to design an efficient algorithm that has reasonable complexity and accuracy and can be used not only for Single-Output systems but also for Multi-Output systems.

This section proposes an approach to obtain the guaranteed state estimation based on the minimization of the P-radius of the zonotopic guaranteed state estimation, offering good trade-off between performance and low complexity when computing an outer approximation of the exact uncertain state set using the zonotope-based procedure proposed by Algorithm 2.2. The complex online computation of the SVD method and the minimization of the volume of the zonotope are replaced by an optimization problem solved off-line. Moreover, this P-radius-based method allows us to guarantee the non-increase of the zonotopic guaranteed state estimation. Before detailing the proposed method, the definition of the P-radius is presented as follows.

DEFINITION 2.3.– *The P-radius of a zonotope $\mathcal{Z} = \boldsymbol{p} \oplus H\boldsymbol{B}^m$ is defined by the following expression:*

$$l = \max_{\boldsymbol{z} \in \mathcal{Z}} (\|\boldsymbol{z} - \boldsymbol{p}\|_P^2) \qquad [2.38]$$

where P is a symmetric and positive definite matrix ($P = P^T \succ 0$).

This definition provides a new criterion for characterizing the size of a zonotope with respect to the P matrix and it can be used to value the quality of the estimation by zonotopes. A small value of the P-radius leads to a small zonotope, which signifies a good quality estimation.

The P-radius definition is illustrated in Figure 2.2. This figure shows a centered zonotope $\mathcal{Z}_1 = \mathbf{p}_1 \oplus H_1\mathbf{B}^3 \in \mathbb{R}^2$ constructed by a linear image of a centered cube, with $\mathbf{p}_1 = \begin{bmatrix} 0 \\ 0 \end{bmatrix}$ and $H_1 = \begin{bmatrix} 1 & 2 & 3 \\ 3 & 2 & 1 \end{bmatrix}$, and a centered zonotope

$\mathcal{Z}_2 = \mathbf{p}_2 \oplus H_2\mathbf{B}^3 \in \mathbb{R}^2$ constructed by $\mathbf{p}_2 = \begin{bmatrix} 0 \\ 0 \end{bmatrix}$ and $H_2 = \begin{bmatrix} 1 & 0.4 & 3 \\ 3 & 0.2 & 1 \end{bmatrix}$. The associated P-radius of these two zonotopes are:

$$\begin{aligned} l_1 &= \max_{\mathbf{z}\in\mathcal{Z}_1}(\|\mathbf{z}\|_P^2) = 6^2 + 6^2 = 72 \\ l_2 &= \max_{\mathbf{z}\in\mathcal{Z}_2}(\|\mathbf{z}\|_P^2) = 4.4^2 + 4.2^2 = 37 \end{aligned} \quad [2.39]$$

with the identity matrix $P = \begin{bmatrix} 1 & 0 \\ 0 & 1 \end{bmatrix}$. The associated P-radius of $\mathcal{Z}_1$ is related to the ellipsoid $\mathcal{E}_1$ defined by $\mathbf{x}^T P\mathbf{x} \leq l_1$, and the associated P-radius of $\mathcal{Z}_2$ is related to the ellipsoid $\mathcal{E}_2$ defined by $\mathbf{x}^T P\mathbf{x} \leq l_2$. From this figure, it can be seen that if the zonotope is large, then the value of its P-radius and its related ellipsoid are large and vice versa. The latter is introduced to characterize the zonotope size by the associated P-radius.

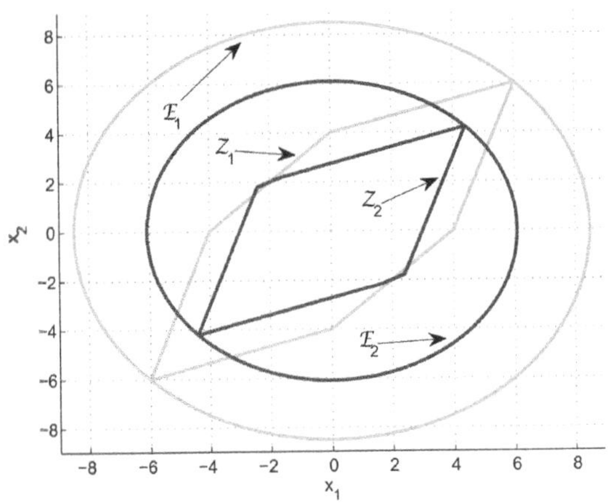

Figure 2.2. *Zonotopes and ellipsoids related to the associated P-radius. For a color version of this figure, see www.iste.co.uk / stoica / zonotopes.zip*

To better emphasize its advantages, this method is detailed in the next chapter both for Single-Output systems and Multi-Output systems.

Chapter 3

Zonotopic Guaranteed State Estimation Based on P-Radius Minimization

This chapter details the P-radius-based procedure for zonotopic guaranteed set-membership estimation for both Single-Output systems and for Multi-Output systems.

3.1. Single-Output systems approach

The P-radius-based method is developed first for linear discrete-time invariant Single-Output systems [1]:

$$\begin{cases} \mathbf{x}_{k+1} = A\mathbf{x}_k + \omega_k \\ y_k = \mathbf{c}^T\mathbf{x}_k + v_k \end{cases} \qquad [3.1]$$

The disturbance is bounded by a centered zonotope $\omega_k \in \mathcal{W} = F\mathbf{B}^{n_x}$ and the measurement noise is bounded by a

1 This is a particular case of [2.5].

centered interval $v_k \in \mathcal{V} = \sigma\mathbf{B}^1 \subset \mathbb{R}$ with $\sigma \in \mathbb{R}^+$. The aim is to find a zonotopic guaranteed state estimation at each sample time. Similar to the optimization-based methods presented in the previous chapter, this set is the outer approximation of the intersection of the two sets $\mathcal{X}_{\mathbf{y}_k}$ and $\bar{\mathcal{X}}_k$. With the definition of $\mathcal{V}$, the measurement consistent state set at time k is defined as a strip: $\mathcal{X}_{\mathbf{y}_k} = \{\mathbf{x} \in \mathbb{R}^n : |\mathbf{c}^T\mathbf{x} - y_k| \leq \sigma\}$. As $\bar{\mathcal{X}}_k$ is a zonotope and $\mathcal{X}_{\mathbf{y}_k}$ is a strip, a family of zonotopes (parameterized by a vector $\boldsymbol{\lambda}$), which contains the intersection between $\bar{\mathcal{X}}_k$ and $\mathcal{X}_{\mathbf{y}_k}$, is computed using Proposition 2.1. The idea is to find the value of this vector $\boldsymbol{\lambda}$ that minimizes the size of the zonotopic approximation based on the P-radius minimization criterion in order to find a good trade-off between the previous zonotopic set-membership estimation methods presented in the previous chapter.

Supposing an outer approximation of the state set $\hat{\mathcal{X}}_{k-1} = \hat{\mathbf{p}}_{k-1} \oplus \hat{H}_{k-1}\mathbf{B}^r$ at time instant $k-1$ and the measured output y_k at time instant k, the predicted state set $\bar{\mathcal{X}}_k$ and the guaranteed state estimation $\hat{\mathcal{X}}_k(\boldsymbol{\lambda})$ are obtained similar to the previous two optimization-based methods (i.e. the zonotope segments minimization method and the zonotope volume minimization method):

$$\bar{\mathcal{X}}_k = A\hat{\mathbf{p}}_{k-1} \oplus \left[A\hat{H}_{k-1}\ F\right]\mathbf{B}^{r+n_x} = \bar{\mathbf{p}}_k \oplus \bar{H}_k\mathbf{B}^{r+n_x} \qquad [3.2]$$

$$\hat{\mathcal{X}}_k(\boldsymbol{\lambda}) = \hat{\mathbf{p}}_k(\boldsymbol{\lambda}) \oplus \hat{H}_k(\boldsymbol{\lambda})\mathbf{B}^{r+n_x+1} \qquad [3.3]$$

with $\hat{\mathbf{p}}_k(\boldsymbol{\lambda}) = A\hat{\mathbf{p}}_{k-1} + \boldsymbol{\lambda}(y_k - \mathbf{c}^T A\hat{\mathbf{p}}_{k-1})$ and $\hat{H}_k(\boldsymbol{\lambda}) = \left[(I - \boldsymbol{\lambda}\mathbf{c}^T)A\hat{H}_{k-1}\ (I - \boldsymbol{\lambda}\mathbf{c}^T)F\ \sigma\boldsymbol{\lambda}\right]$

Denote the P-radius of the zonotopic state estimation set at time instant k by:

$$l_k = \max_{\mathbf{x}\in\hat{\mathcal{X}}_k}(\|\mathbf{x} - \hat{\mathbf{p}}_k\|_P^2) \qquad [3.4]$$

From the definition of the guaranteed state estimation $\hat{\mathcal{X}}_k$ in [3.3], the P-radius can be rewritten as $l_k = \max_{\hat{\mathbf{z}} \in \mathbf{B}^{r+n_x+1}} \|\hat{H}_k(\boldsymbol{\lambda})\hat{\mathbf{z}}\|_P^2$. In this context, the following proposition formulates an interesting result.

PROPOSITION 3.1.– A symmetric positive definite matrix $P = P^T \succ 0$ and a vector $\boldsymbol{\lambda}$ can be computed such that at each sample time the P-radius of the zonotopic state estimation set $\hat{X}_k$ is not increased, more precisely the value of l_k. This means that the zonotopic state estimation set is non-increasing in time.

This proposition on the non-increasing condition of the P-radius can be visualized in Figure 3.1 where the zonotope is the guaranteed state estimation at each iteration and the ellipsoid $\|\mathbf{x} - \hat{\mathbf{p}}_k\|_P^2 \leq l_k$ is related to the P-radius of the zonotopic state estimation set $\hat{\mathcal{X}}_k$. The reader should not fear the fact that the zonotope is partially out of the ellipsoid because this ellipsoid is only a criterion to characterize the size of the zonotope.

A constructive proof of Proposition 3.1 is presented in the following. The non-increasing condition of the P-radius can be expressed by a mathematical formula as follows. The contractiveness of the P-radius l_k is ensured by the expression $l_k \leq \beta l_{k-1}$, with $\beta \in (0, 1]$. Because of the presence of disturbances and measurement noise, this condition is difficult to verify. A relaxation of this condition can be $l_k \leq \beta l_{k-1} + \epsilon$, where ϵ is a positive constant which allows us to bound the influence of disturbances and measurement noises. For $\epsilon = \max_{\mathbf{s} \in \mathbf{B}^{n_x}} \|F\mathbf{s}\|_2^2 + \sigma^2$, this leads to the following inequality:

$$l_k \leq \beta l_{k-1} + \max_{\mathbf{s} \in \mathbf{B}^{n_x}} \|F\mathbf{s}\|_2^2 + \sigma^2 \qquad [3.5]$$

with $\beta \in (0, 1]$ and $\max_{\mathbf{s} \in \mathbf{B}^{n_x}} \|F\mathbf{s}\|_2^2 + \sigma^2 > 0$.

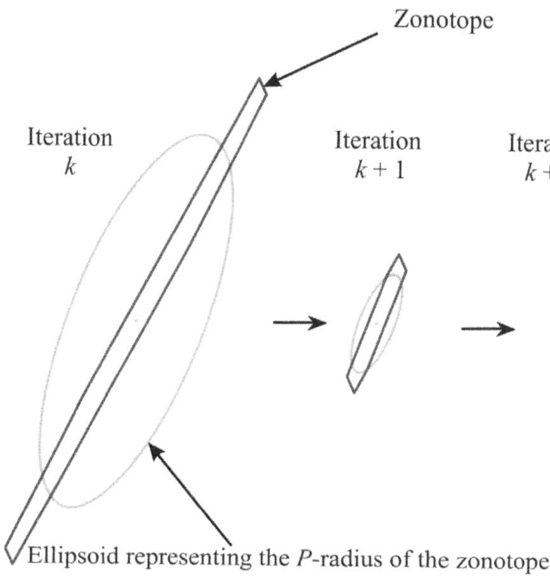

Figure 3.1. *Evolution of the guaranteed state estimation*

This inequality can be rewritten in an equivalent form:

$$\max_{\hat{\mathbf{z}}\in\mathbf{B}^{r+n_x+1}} \|\hat{H}_k(\boldsymbol{\lambda})\hat{\mathbf{z}}\|_P^2 \leq \max_{\mathbf{z}\in\mathbf{B}^r} \beta\|\hat{H}_{k-1}\mathbf{z}\|_P^2 + \max_{\mathbf{s}\in\mathbf{B}^{n_x}} \|F\mathbf{s}\|_2^2 + \sigma^2 \quad [3.6]$$

with $\hat{\mathbf{z}} = \left[\mathbf{z}^T \ \mathbf{s}^T \ \eta^T\right]^T \in \mathbf{B}^{r+n_x+1}$, $\eta \in \mathbf{B}^1$, and $\beta \in (0, 1]$.

The following inequality leads to a sufficient condition for the previous expression [3.6]:

$$\max_{\hat{\mathbf{z}}\in\mathbf{B}^{r+n_x+1}} (\|\hat{H}_k(\boldsymbol{\lambda})\hat{\mathbf{z}}\|_P^2 - \beta\|\hat{H}_{k-1}\mathbf{z}\|_P^2 - \|F\mathbf{s}\|_2^2 - \sigma^2) \leq 0 \qquad [3.7]$$

This is equivalent to the following inequality:

$$\hat{\mathbf{z}}^T \hat{H}_k^T(\boldsymbol{\lambda}) P \hat{H}_k(\boldsymbol{\lambda})\hat{\mathbf{z}} - \beta\mathbf{z}^T \hat{H}_{k-1}^T P \hat{H}_{k-1}\mathbf{z} - \mathbf{s}^T F^T F\mathbf{s} - \sigma^2 \leq 0, \quad \forall \mathbf{z}, \mathbf{s}, \eta \qquad [3.8]$$

Because $\eta \in \mathbf{B}^1$, which is equivalent to $|\eta| \leq 1$, the following expression is obtained:

$$\sigma^2(1-\eta^2) \geq 0 \qquad [3.9]$$

Adding this positive term to the left side of [3.8] leads to the following sufficient condition for [3.8]:

$$\begin{aligned}&\hat{\mathbf{z}}^T \hat{H}_k^T(\boldsymbol{\lambda}) P \hat{H}_k(\boldsymbol{\lambda})\hat{\mathbf{z}} - \beta \mathbf{z}^T \hat{H}_{k-1}^T P \hat{H}_{k-1}\mathbf{z} - \mathbf{s}^T F^T F \mathbf{s} - \sigma^2 + \\ &+\sigma^2(1-\eta^2) \leq 0, \quad \forall \mathbf{z}, \mathbf{s}, \eta\end{aligned} \qquad [3.10]$$

Right multiplying the expression of $\hat{H}_k$ in [3.3] with the explicit form of $\hat{\mathbf{z}} = [\mathbf{z}^T \mathbf{s}^T \eta^T]^T$ leads to:

$$\hat{H}_k(\boldsymbol{\lambda})\hat{\mathbf{z}} = (I - \boldsymbol{\lambda}\mathbf{c}^T)(A\hat{H}_{k-1}\mathbf{z} + F\mathbf{s}) + \sigma\boldsymbol{\lambda}\eta \qquad [3.11]$$

Denote $\boldsymbol{\theta} = \hat{H}_{k-1}\mathbf{z}$, then inequality [3.10] can be written in the following matrix form:

$$\begin{bmatrix}\boldsymbol{\theta} \\ \mathbf{s} \\ \eta\end{bmatrix}^T \begin{bmatrix} A_{11} & A_{12} & A_{13} \\ * & A_{22} & A_{23} \\ * & * & A_{33}\end{bmatrix} \begin{bmatrix}\boldsymbol{\theta} \\ \mathbf{s} \\ \eta\end{bmatrix} \leq 0, \quad \forall \boldsymbol{\theta}, \mathbf{s}, \eta \qquad [3.12]$$

with "$*$" denoting the terms required for the symmetry of the matrix and the following additional notations:

$$\begin{cases} A_{11} = ((I - \boldsymbol{\lambda}\mathbf{c}^T)A)^T P (I - \boldsymbol{\lambda}\mathbf{c}^T)A - \beta P \\ A_{12} = ((I - \boldsymbol{\lambda}\mathbf{c}^T)A)^T P (I - \boldsymbol{\lambda}\mathbf{c}^T)F \\ A_{13} = ((I - \boldsymbol{\lambda}\mathbf{c}^T)A)^T P \sigma \boldsymbol{\lambda} \\ A_{22} = ((I - \boldsymbol{\lambda}\mathbf{c}^T)F)^T P (I - \boldsymbol{\lambda}\mathbf{c}^T)F - F^T F \\ A_{23} = ((I - \boldsymbol{\lambda}\mathbf{c}^T)F)^T P \sigma \boldsymbol{\lambda} \\ A_{33} = \sigma^2 \boldsymbol{\lambda}^T P \boldsymbol{\lambda} - \sigma^2. \end{cases} \qquad [3.13]$$

Using the definition of positive definite matrix (see Appendix 1) allows us to rewrite [3.12] as:

$$\begin{bmatrix} A_{11} & A_{12} & A_{13} \\ * & A_{22} & A_{23} \\ * & * & A_{33}\end{bmatrix} \preceq 0, \quad \forall \begin{bmatrix}\boldsymbol{\theta} \\ \mathbf{s} \\ \eta\end{bmatrix} \neq 0 \qquad [3.14]$$

or equivalent to:

$$\begin{bmatrix} -A_{11} & -A_{12} & -A_{13} \\ * & -A_{22} & -A_{23} \\ * & * & -A_{33} \end{bmatrix} \succeq 0, \quad \forall \begin{bmatrix} \boldsymbol{\theta} \\ \mathbf{s} \\ \eta \end{bmatrix} \neq 0 \qquad [3.15]$$

With the explicit notations [3.13], expression [3.15] can be rewritten in a more appropriate form leading to the following matrix inequality:

$$\begin{bmatrix} \beta P & 0 & 0 \\ * & F^T F & 0 \\ * & * & \sigma^2 \end{bmatrix} -$$

$$\begin{bmatrix} (A^T - A^T \mathbf{c} \boldsymbol{\lambda}^T) P \\ (F^T - F^T \mathbf{c} \boldsymbol{\lambda}^T) P \\ \boldsymbol{\lambda}^T P \sigma \end{bmatrix} P^{-1} \begin{bmatrix} (A^T - A^T \mathbf{c} \boldsymbol{\lambda}^T) P \\ (F^T - F^T \mathbf{c} \boldsymbol{\lambda}^T) P \\ \boldsymbol{\lambda}^T P \sigma \end{bmatrix}^T \succeq 0 \quad [3.16]$$

Using the Schur complement definition (see Appendix 1), expression [3.16] is equivalent to the following matrix inequality:

$$\begin{bmatrix} \beta P & 0 & 0 & A^T P - A^T \mathbf{c} Y^T \\ * & F^T F & 0 & F^T P - F^T \mathbf{c} Y^T \\ * & * & \sigma^2 & Y^T \sigma \\ * & * & * & P \end{bmatrix} \succeq 0 \qquad [3.17]$$

with a change of variables $Y = P\boldsymbol{\lambda}$, with $P \in \mathbb{R}^{n_x \times n_x}$, $\boldsymbol{\lambda} \in \mathbb{R}^{n_x}$ and $Y \in \mathbb{R}^{n_x}$.

As the 2-norm is a convex function and $\mathcal{W}$ is a convex set, the term $\max_{\mathbf{s} \in \mathbf{B}^{n_x}} \|F\mathbf{s}\|_2^2$ can be easily computed using the maximum principle [ROC 70]. Thus, the value of $\epsilon = \max_{\mathbf{s} \in \mathbf{B}^{n_x}} \|F\mathbf{s}\|_2^2 + \sigma^2$ can be obtained. Then condition [3.5] can be written as $l_k \leq \beta l_{k-1} + \epsilon$. At infinity, this expression is equivalent to:

$$l_\infty = \beta l_\infty + \epsilon \qquad [3.18]$$

leading to:

$$l_\infty = \frac{\epsilon}{1-\beta} \qquad [3.19]$$

Let us consider an ellipsoid $\mathcal{E} = \{\mathbf{x} : \mathbf{x}^T P \mathbf{x} \le \frac{\epsilon}{1-\beta}\}$ which can be normalized to $\mathcal{E} = \{\mathbf{x} : \mathbf{x}^T \frac{(1-\beta)P}{\epsilon} \mathbf{x} \le 1\}$. This ellipsoid is related to the P-radius of the zonotopic guaranteed state estimation at infinity. To minimize the P-radius (i.e. l_∞) of the zonotope, we can find the ellipsoid of smallest diameter [BOY 94]. This leads to solving the following Eigenvalue Problem (EVP):

$$\max_{\tau,\beta,P} \tau$$

subject to the BMI

$$\frac{(1-\beta)P}{\epsilon} \succeq \tau I_{n_x} \qquad [3.20]$$

where $I_{n_x} \in \mathbb{R}^{n_x \times n_x}$ is the identity matrix, $\tau \in \mathbb{R}^+$ and $\beta \in (0,1]$ are scalars. Then the smallest diameter is computed by $\frac{2}{\sqrt{\tau^*}}$ [BOY 94].

Finally, to find the values of $P = P^T \succ 0 \in \mathbb{R}^{n_x \times n_x}$ and $\boldsymbol{\lambda} \in \mathbb{R}^{n_x}$, the following optimization problem must be solved:

$$\max_{\tau,\beta,P,Y} \tau$$

subject to

$$\begin{cases} \frac{(1-\beta)P}{\epsilon} \succeq \tau I \\ \begin{bmatrix} \beta P & 0 & 0 & A^T P - A^T \mathbf{c} Y^T \\ * & F^T F & 0 & F^T P - F^T \mathbf{c} Y^T \\ * & * & \sigma^2 & Y^T \sigma \\ * & * & * & P \end{bmatrix} \succeq 0 \\ \tau > 0 \end{cases} \qquad [3.21]$$

As β is a scalar variable, this optimization problem can be efficiently solved by using a Bilinear Matrix Inequality (BMI) solver (e.g. *Penbmi* [KOČ 03]) or by executing a simple search-loop on β leading to a Linear Matrix Inequality (LMI) problem. In this optimization problem [3.21], the decision variables are $P = P^T \in \mathbb{R}^{n_x \times n_x}$, $Y \in \mathbb{R}^{n_x}$, $\beta \in (0, 1]$ and $\tau \in \mathbb{R}^+$. Thus, the total number of scalar decision variables is $\frac{n_x(n_x+1)}{2} + n_x + 2$. The dimensions of the inequalities in [3.21] are $n_x \times n_x$, $(3n_x + 1) \times (3n_x + 1)$ and 1, respectively.

In [LE 11b], a modification of problem [3.21] to avoid solving a BMI optimization problem is presented. Instead of optimizing the value of the P-radius, a minimum value of β is searched allowing a maximum decreasing speed of the P-radius. This criterion leads to a new LMI optimization problem based on the bisection algorithm [BUR 00] on β:

$$\min_{\beta \in (0,1]} \beta$$

such that

$$\max_{\tau,P,Y} \tau$$

subject to

$$\begin{cases} \frac{(1-\beta)P}{\epsilon} \succeq \tau I \\ \begin{bmatrix} \beta P & 0 & 0 & A^T P - A^T \mathbf{c} Y^T \\ * & F^T F & 0 & F^T P - F^T \mathbf{c} Y^T \\ * & * & \sigma^2 & Y^T \sigma \\ * & * & * & P \end{bmatrix} \succeq 0 \\ \tau > 0 \end{cases} \qquad [3.22]$$

As β is computed by a bisection algorithm, it is not a decision variable in [3.22]. This means that, in this case, the matrix inequalities [3.22] are LMIs.

To better understand the proposed methods, an illustrative example is further presented.

EXAMPLE 3.1.– Consider the following linear discrete-time invariant system:

$$\begin{cases} \mathbf{x}_{k+1} = \begin{bmatrix} 1 & 1 \\ 0 & 1 \end{bmatrix} \mathbf{x}_k + 0.02 \begin{bmatrix} -6 \\ 1 \end{bmatrix} \omega_k \\ y_k = \begin{bmatrix} -2 & 1 \end{bmatrix} \mathbf{x}_k + 0.2 v_k \end{cases} \quad [3.23]$$

with $\|v_k\|_\infty \leq 1$, $\|\omega_k\|_\infty \leq 1$. The values of v_k and ω_k are generated by random functions with *Matlab*®. The initial state belongs to the box $3\mathbf{B}^2$ (see the solid-line rectangle in Figure 3.2). The guaranteed state estimation must be determined at each time instant. The order of the m-zonotopes in each example in this chapter is limited to $m \leq 20$ in the interest of a fast simulation.

The over-approximation of a high-order zonotope by a low-order zonotope is made using Proposition 1.4 with a Euclidean norm-based criterion and complexity reduction of zonotopes via the interval hull method (see Proposition 1.1). A system of order 2 is chosen to reduce the complexity of the computation and to facilitate the graphical visualization.

The BMI optimization problem [3.21] is solved by two different solvers: the *Penbmi* solver and the LMI solver (*mincx*) of *Matlab*® with a search loop on β (the step on β is 10^{-4}, i.e. $\beta = 10^{-4} : 10^{-4} : 1$), which in this case give the same results: $\beta = 0.4090$, $\boldsymbol{\lambda} = \begin{bmatrix} -0.6205 \\ -0.2842 \end{bmatrix}$. The solution of the optimization problem [3.22] gives $\beta = 0.0001$, $\boldsymbol{\lambda} = \begin{bmatrix} -0.7500 \\ -0.4999 \end{bmatrix}$. The number of scalar decision variables in problems [3.21] and [3.22] is 7 and 6, respectively. The dimensions of these matrix inequalities is 2×2, 7×7 and 1, respectively. The estimation accuracy of the optimization problems [3.21] and [3.22] is compared with the results obtained from the minimization of the segments of a zonotope method and the minimization of the volume of a zonotope method.

Figures 3.2 and 3.3 show the evolution of the predicted state set and the outer approximation of state estimation set using the optimization problems [3.21] and [3.22], respectively. The outer approximations are rapidly reduced at each iteration. This reduction means that the guaranteed state estimation set (an approximation of the real intersection) becomes more and more accurate at each iteration ($k = 1$, $k = 2$, $k = 3$). Comparision of these figures shows that the guaranteed state estimation obtained by the optimization problem [3.22] is decreased more rapidly than the guaranteed state estimation obtained by the optimization problem [3.21] due to the minimum value of β obtained via [3.22].

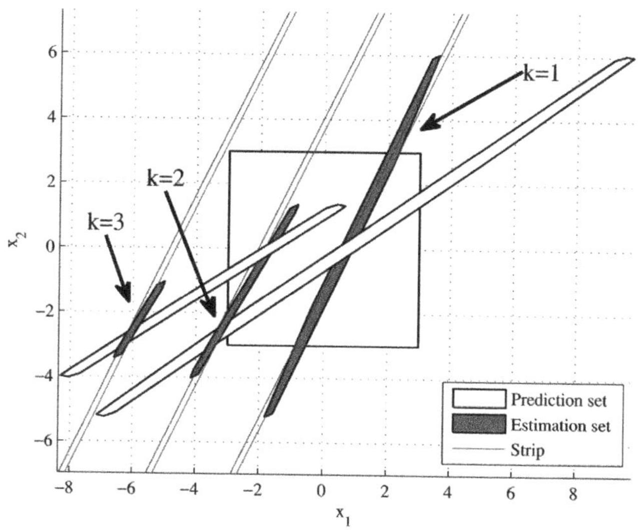

Figure 3.2. *Evolution of the guaranteed state estimation via the optimization problem [3.21] (Example 3.1)*

Figure 3.4 and its zoom (Figure 3.5) show the bounds on x_{2_k} obtained by the optimization problem [3.21], the optimization problem [3.22] and the two methods developed in [ALA 05]. The dashed lines represent the bounds of x_{2_k}

obtained by the segment minimization algorithm; the dashed-dotted lines represent the bounds of x_{2_k} obtained by the volume minimization algorithm; the solid lines represent the bounds of x_{2_k} obtained by the optimization problem [3.21]; the dotted lines represent the bounds of x_{2_k} obtained by the optimization problem [3.22]; and the stars represent the real state x_{2_k} of the system. These points are found inside the bounds of x_{2_k} confirming that this bound is well estimated by each method. To compare the accuracy of these methods, the bound's width of states will be analyzed in the next figures.

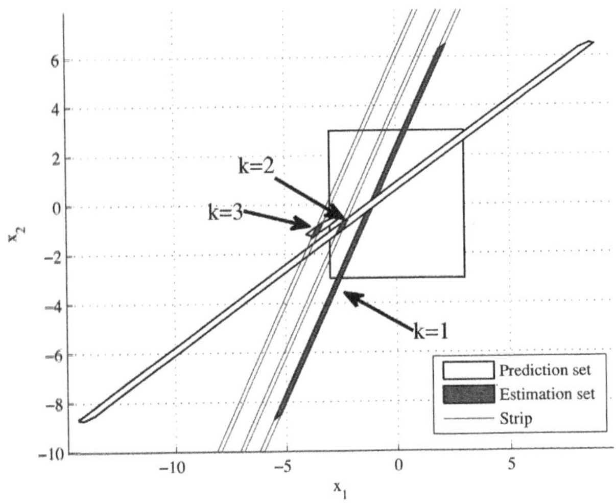

Figure 3.3. *Evolution of the guaranteed state estimation via the optimization problem [3.22] (Example 3.1)*

To confirm the good accuracy of the P-radius-based approach, Figure 3.8 compares the volume of the zonotopic guaranteed state estimation by different approaches (the segment minimization, the volume minimization and the P-radius minimization). The P-radius minimization gives better accuracy than the segment minimization and as good as the volume minimization.

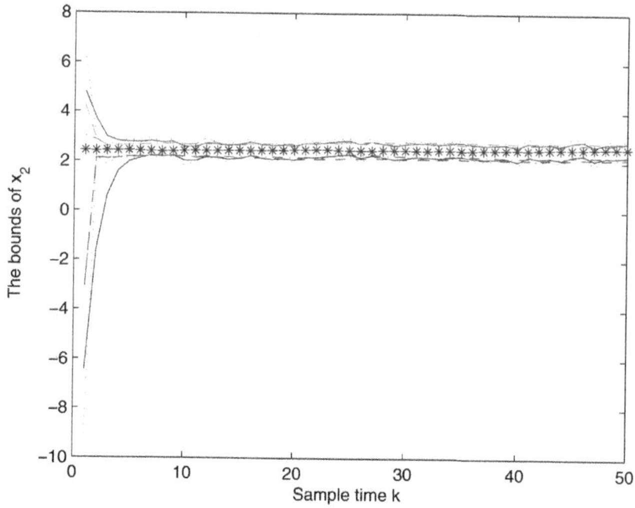

Figure 3.4. *Bounds of x_2 via different approaches (Example 3.1)*

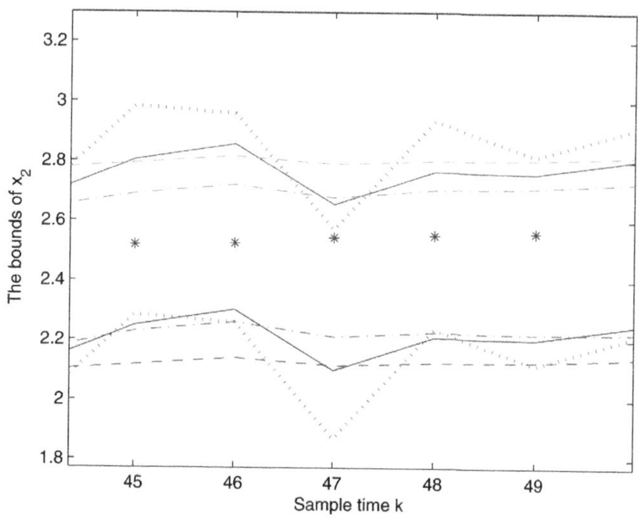

Figure 3.5. *Zoom of the bounds of x_2 via different approaches (Example 3.1)*

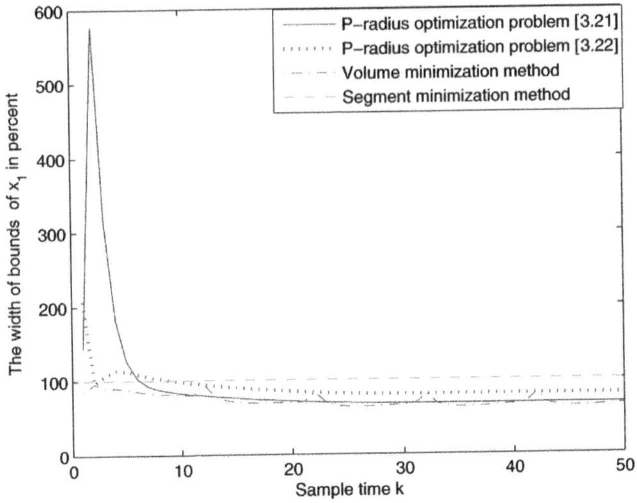

Figure 3.6. *Comparison of the bound's width of x_1 via different approaches (Example 3.1)*

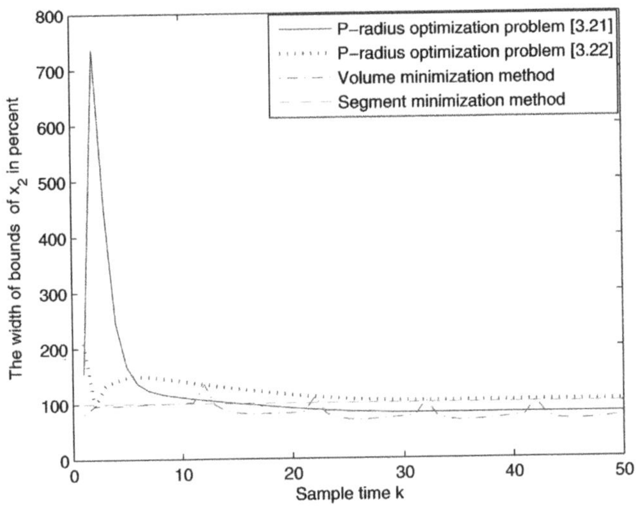

Figure 3.7. *Comparison of the bound's width of x_2 via different approaches (Example 3.1)*

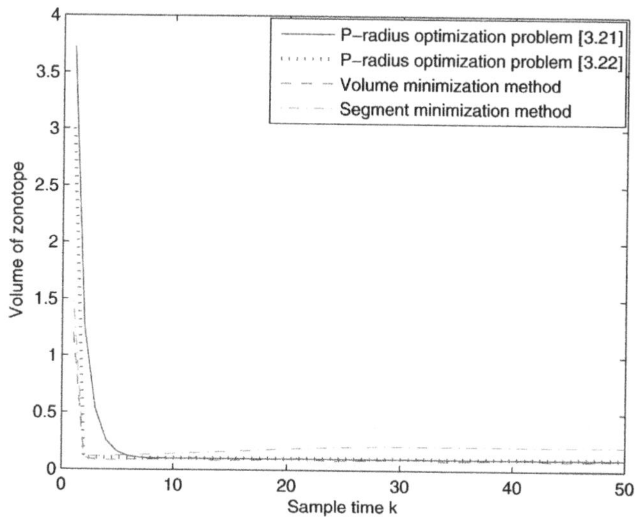

Figure 3.8. *Comparison of the volume of guaranteed state estimation via different approaches (Example 3.1)*

Algorithm	Time (seconds)
Segment minimization	0.0468
P-radius optimization problem (without off-line optimization problem included)	0.0312
P-radius optimization problem (with off-line optimization [3.21] included)	0.9204
P-radius optimization problem (with off-line optimization [3.22] included)	0.6240
Volume minimization	10.0153

Table 3.1. *Total computation time of Example 3.1 after 50 time instants*

Table 3.1 shows the computation time of different algorithms. These results are obtained with an Intel Core $i5$ 2.67 GHz. The BMI optimization [3.21] is solved by the *Penbmi* solver; the LMI optimization [3.22] is dealt with by LMI toolbox ($mincx$) of $Matlab^{®}$; the volume minimization problem is solved by the $fminsearch$ function of Matlab; and the segment minimization problem is solved with a simple matrix computation. If the time used to solve LMI optimizations is not taken into account, the computation time of this method is the same as the computation of the segment minimization but the accuracy of the estimation is better.

Even if the LMI optimizations are taken into account, the computation time is 10 times less than the computation time of the volume minimization method. This highlights the advantages of the proposed method on the computation complexity.

The segment minimization algorithm has an acceptable accuracy and a very short computation time (only a few computations to obtain λ). The volume minimization algorithm gives better accuracy, but it needs a longer computation time (215 times more than for the segment minimization method) because an optimization problem must be solved at each sample time. The accuracy of the presented method can be comparable with the accuracy obtained by the volume minimization algorithm, but λ is computed at the beginning of the program and at each iteration the value of λ does not need to be recomputed leading to a reasonable online complexity. In summary, the proposed P-radius-based algorithms combine the advantage of the volume minimization (accuracy) and the segment minimization (computation time). In addition, the P-radius-based approaches allow us to overcome the problem of volume minimization due to a very narrow zonotope.

3.2. Multi-Output systems approaches

The presented estimation methodology is further extended to the case of Multi-Output systems. The diagram proposed in Figure 3.9 recapitulates the solution of the estimation problem for Multi-Output systems. First, a general procedure for Multi-Output systems is described, allowing us to introduce new useful notations. Second, direct extensions (based on the intersection between a zonotope and a strip) of Single-Output systems to the case of Multi-Output systems are proposed. Third, an improved solution (based on the intersection between a zonotope and a polytope

[LE 12b, LE 13]) developed directly for Multi-Output systems is presented.

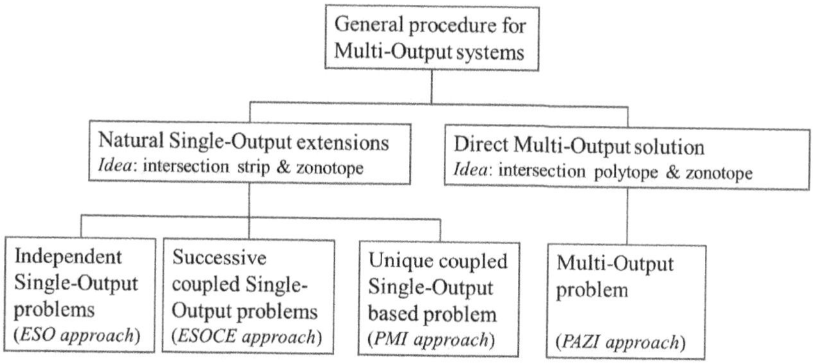

Figure 3.9. *Classification of proposed solution for Multi-Output systems*

3.2.1. *General formulation*

Considering the Multi-Output system [2.5], the guaranteed state estimation set $\hat{\mathcal{X}}_k$ can be found by successively repeating the zonotopic guaranteed state estimation approach presented in the previous section for each component of the output measurement vector $\mathbf{y}_k$, denoted $y_{k/i}$:

$$y_{k/i} = \mathbf{c}_i^T \mathbf{x}_k + v_{k/i}, i = 1, \ldots, n_y \qquad [3.24]$$

where $\mathbf{c}_i^T$ is the ith row of matrix C and the noise $v_{k/i}$ is bounded by the interval $\mathcal{V}_i = \sigma_i \mathbf{B}^1$, with $\sigma_i = \Sigma_{ii}$. Thus, the measurement noise v_k belongs to a box $\mathcal{V} = diag(\sigma_1, \ldots, \sigma_{n_y})\mathbf{B}^{n_y}$. If this assumption is not satisfied, an outer approximation of $\mathcal{V}$ by a box can be used.

Supposing an outer approximation of the state set $\hat{\mathcal{X}}_{k-1} = \hat{\mathbf{p}}_{k-1} \oplus \hat{H}_{k-1}\mathbf{B}^r$ at time instant $k-1$, then the predicted state set at the next time instant $\bar{\mathcal{X}}_k$ can be computed as in [3.2]. The exact estimation set will be obtained after intersecting the predicted state set with the measurement consistent state set of the measured output vector $\mathbf{y}_k$ (as described in Algorithm 2.2). In a general way, the outer approximation of this set can be found by using the guaranteed state intersection as follows.

Similar to [3.3], an outer approximation of the intersection between the strip created by the first component of the output vector $y_{k/1}$ and the predicted state set $\bar{\mathcal{X}}_k$ is represented by:

$$\hat{\mathcal{X}}_{k/1}(\boldsymbol{\lambda}_1) = \hat{\mathbf{p}}_{k/1}(\boldsymbol{\lambda}_1) \oplus \hat{H}_{k/1}(\boldsymbol{\lambda})\mathbf{B}^{r+n_x+1} \qquad [3.25]$$

where $\hat{\mathbf{p}}_{k/1}(\boldsymbol{\lambda}_1) = A\hat{\mathbf{p}}_{k-1} + \boldsymbol{\lambda}_1(y_{k/1} - \mathbf{c}_1^T A\hat{\mathbf{p}}_{k-1})$
and $\hat{H}_{k/1}(\boldsymbol{\lambda}_1) = \left[(I - \boldsymbol{\lambda}_1\mathbf{c}_1^T)A\hat{H}_{k-1}\; (I - \boldsymbol{\lambda}_1\mathbf{c}_1^T)F\; \sigma_1\boldsymbol{\lambda}_1\right]$

Then this set $\hat{\mathcal{X}}_{k/1}(\boldsymbol{\lambda}_1)$ is intersected with the strip obtained from the second component of the measured output vector ($y_{k/2}$) leading to:

$$\hat{\mathcal{X}}_{k/2}(\boldsymbol{\lambda}_1, \boldsymbol{\lambda}_2) = \hat{\mathbf{p}}_{k/2}(\boldsymbol{\lambda}_1, \boldsymbol{\lambda}_2) \oplus \hat{H}_{k/2}(\boldsymbol{\lambda}_1, \boldsymbol{\lambda}_2)\mathbf{B}^{r+n_x+2} \qquad [3.26]$$

with $\hat{\mathbf{p}}_{k/2}(\boldsymbol{\lambda}_1, \boldsymbol{\lambda}_2) = \hat{\mathbf{p}}_{k/1}(\boldsymbol{\lambda}_1) + \boldsymbol{\lambda}_2(y_{k/2} - \mathbf{c}_2^T\hat{\mathbf{p}}_{k/1}(\boldsymbol{\lambda}_1))$
and $\hat{H}_{k/2}(\boldsymbol{\lambda}_1, \boldsymbol{\lambda}_2) = \left[(I - \boldsymbol{\lambda}_2\mathbf{c}_2^T)\hat{H}_{k/1}(\boldsymbol{\lambda}_1)\; \sigma_2\boldsymbol{\lambda}_2\right]$

This procedure is repeated until the last component of the measured output vector (y_{k/n_y}) leading to the zonotopic guaranteed state estimation set at time instant k:

$$\begin{aligned}\hat{\mathcal{X}}_{k/n_y}(\boldsymbol{\lambda}_1, \ldots, \boldsymbol{\lambda}_{n_y}) = \hat{\mathbf{p}}_{k/n_y}(\boldsymbol{\lambda}_1, \ldots, \boldsymbol{\lambda}_{n_y}) \oplus \\ \oplus \hat{H}_{k/n_y}(\boldsymbol{\lambda}_1, \ldots, \boldsymbol{\lambda}_{n_y})\mathbf{B}^{r+n_x+n_y}\end{aligned} \qquad [3.27]$$

with the recursive notations:

$$\hat{\mathbf{p}}_{k/n_y}(\boldsymbol{\lambda}_1,\ldots,\boldsymbol{\lambda}_{n_y}) = \hat{\mathbf{p}}_{k/n_y-1}(\boldsymbol{\lambda}_1,\ldots,\boldsymbol{\lambda}_{n_y-1}) + \boldsymbol{\lambda}_{n_y}(y_{k/n_y} - \mathbf{c}_{n_y}^T\hat{\mathbf{p}}_{k/n_y-1}(\boldsymbol{\lambda}_1,\ldots,\boldsymbol{\lambda}_{n_y-1})) \quad [3.28]$$

and

$$\hat{H}_{k/n_y}(\boldsymbol{\lambda}_1,\ldots,\boldsymbol{\lambda}_{n_y}) = \left[(I - \boldsymbol{\lambda}_{n_y}\mathbf{c}_{n_y}^T)\hat{H}_{k/n_y-1}(\boldsymbol{\lambda}_1,\ldots,\boldsymbol{\lambda}_{n_y-1}) \;\; \sigma_{n_y}\boldsymbol{\lambda}_{n_y}\right] \quad [3.29]$$

Finally, the zonotopic guaranteed state estimation set at time instant k is provided by:

$$\hat{\mathcal{X}}_k(\boldsymbol{\lambda}_1,\ldots,\boldsymbol{\lambda}_{n_y}) = \hat{\mathbf{p}}_k(\boldsymbol{\lambda}_1,\ldots,\boldsymbol{\lambda}_{n_y}) \oplus \hat{H}_k(\boldsymbol{\lambda}_1,\ldots,\boldsymbol{\lambda}_{n_y})\mathbf{B}^{r+n_x+n_y} \quad [3.30]$$

with $\hat{\mathbf{p}}_k(\boldsymbol{\lambda}_1,\ldots,\boldsymbol{\lambda}_{n_y}) = \hat{\mathbf{p}}_{k/n_y}(\boldsymbol{\lambda}_1,\ldots,\boldsymbol{\lambda}_{n_y})$
and $\hat{H}_k(\boldsymbol{\lambda}_1,\ldots,\boldsymbol{\lambda}_{n_y}) = \hat{H}_{k/n_y}(\boldsymbol{\lambda}_1,\ldots,\boldsymbol{\lambda}_{n_y})$

This procedure can be visualized in Figure 3.10 for the case of two outputs. Suppose at time instant k from the guaranteed state estimation set $\hat{\mathcal{X}}_{k-1}$ the predicted state set is determined (represented by the zonotope $\bar{\mathcal{X}}_k$). First, this set is intersected with the first element of the measured output denoted by the strip $|\mathbf{c}_1^T\mathbf{x} - y_{k/1}| \leq \sigma_1$. Then this intersection is approximated by the zonotope $\hat{\mathcal{X}}_{k/1}$ using Proposition 3.1. The procedure is repeated with the second element of the measured output (the strip $|\mathbf{c}_2^T\mathbf{x} - y_{k/2}| \leq \sigma_2$). Proposition 3.1 is used again to obtain the outer approximation of the intersection between $\hat{\mathcal{X}}_{k/1}$ and the second strip. The guaranteed state estimation set is then the zonotope $\hat{\mathcal{X}}_{k/2}$.

Later on, in the general case of n_y outputs, three original procedures are proposed to compute the vectors $\boldsymbol{\lambda}_1,\ldots,\boldsymbol{\lambda}_{n_y}$.

3.2.2. *Extensions of the Single-Output systems methodology*

In order to apply the guaranteed state estimation to Multi-Output systems, three cases are further considered and compared. First, the Multi-Output problem is decoupled in several independent Single-Output problems. Second, successive coupled Single-Output problems are considered. Third, a unique coupled Single-Output-based problem is proposed leading to a Polynomial Matrix Inequality (PMI) problem.

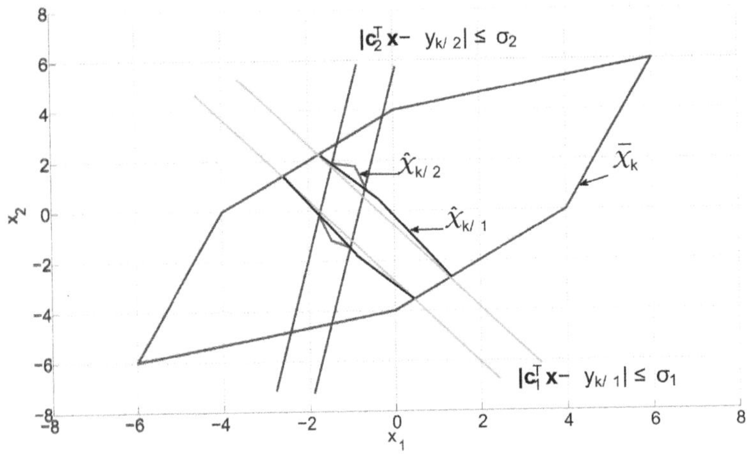

Figure 3.10. *State estimation for a two-output system. For a color version of this figure, see www.iste.co.uk/stoica/zonotopes.zip*

3.2.2.1. *Equivalent Single-Output approach*

In this approach, the Multi-Output system is considered as n_y independent Single-Output systems. This leads to independently computing the vectors $\lambda_1, \ldots, \lambda_{n_y}$ by independently solving n_y separate optimization problems [3.21] or [3.22] off-line. This approach is called the Equivalent Single-Output (ESO) approach. The guaranteed state estimation is determined using the following algorithm:

ALGORITHM 3.1.– ESO

1) For $j = 1, \ldots, n_y$

Step j: Using the strip of the measurement $y_{k/j}$ compute $\boldsymbol{\lambda}_j$;
End.

2) The guaranteed state estimation is computed by equation [3.30] with the known vectors $\boldsymbol{\lambda}_1, \ldots, \boldsymbol{\lambda}_{n_y}$.

This approach proposes a simple solution for the state estimation of Multi-Output systems but this solution can lead to a conservative result due to the possible Multi-Output coupling effect. In the following section, another solution is proposed to improve the accuracy of the guaranteed state estimation.

3.2.2.2. *Equivalent Single-Output with Coupling Effect approach*

Furthermore, to reduce the conservatism of the previous method, the following procedure is proposed. Using $y_{k/1}$, the predicted state set $\bar{\mathcal{X}}_k$ and method [3.21] or method [3.22] leads to $\boldsymbol{\lambda}_1$ and a smaller zonotope $\hat{\mathcal{X}}_{k/1}$. Intersecting this new zonotope with the strip corresponding to $y_{k/2}$ (supposing $\boldsymbol{\lambda}_1$ is known from the previous step) leads to $\boldsymbol{\lambda}_2$ and another zonotope $\hat{\mathcal{X}}_{k/2}$. This procedure is repeated until the last component of the output vector y_{k/n_y} (supposing all the previous vectors $\boldsymbol{\lambda}_1, \ldots, \boldsymbol{\lambda}_{n_y-1}$ are known). This approach is called the Equivalent Single-Output with Coupling Effect (ESOCE) approach. The following algorithm describes this off-line procedure:

ALGORITHM 3.2.– ESOCE

1) Step 1: Using the measurement $y_{k/1}$ and [3.21], compute $\boldsymbol{\lambda}_1$;

2) For $j = 2, \ldots, n_y$

Step j: Using the measurement $y_{k/j}$ and the previous obtained

vectors $\boldsymbol{\lambda}_1, \ldots, \boldsymbol{\lambda}_{j-1}$, compute $\boldsymbol{\lambda}_j$.
End.

The computation of $\boldsymbol{\lambda}_j$ is detailed as follows. The guaranteed state estimation set at Step j is computed as:

$$\hat{\mathcal{X}}_{k/j}(\boldsymbol{\lambda}_1, \ldots, \boldsymbol{\lambda}_j) = \hat{\mathbf{p}}_{k/j}(\boldsymbol{\lambda}_1, \ldots, \boldsymbol{\lambda}_j) \oplus \hat{H}_{k/j}(\boldsymbol{\lambda}_1, \ldots, \boldsymbol{\lambda}_j)\mathbf{B}^{r+n_x+j} \quad [3.31]$$

The non-increasing condition on the P-radius of the zonotopic estimation set is applied leading to:

$$\max_{\hat{\mathbf{z}} \in \mathbf{B}^{r+n_x+j}} \|\hat{H}_{k/j}\hat{\mathbf{z}}\|_P^2 \leq \max_{\mathbf{z} \in \mathbf{B}^r} \beta \|\hat{H}_{k-1}\mathbf{z}\|_P^2 + \max_{\mathbf{s} \in \mathbf{B}^{n_x}} \|F\mathbf{s}\|_2^2 + \sigma_1^2 + \ldots + \sigma_j^2 \quad [3.32]$$

with $\hat{\mathbf{z}} = \begin{bmatrix} \mathbf{z}^T & \mathbf{s}^T & \eta_1 & \ldots & \eta_j \end{bmatrix}^T \in \mathbf{B}^{r+n_x+j}$, $\eta_j \in \mathbf{B}^1$, and $\beta \in (0, 1]$.

A sufficient condition for [3.32] is the following:

$$\max_{\hat{\mathbf{z}} \in \mathbf{B}^{r+n_x+j}} (\|\hat{H}_{k/j}\hat{\mathbf{z}}\|_P^2 - \beta\|\hat{H}_{k-1}\mathbf{z}\|_P^2 - \|F\mathbf{s}\|_2^2 - \sigma_1^2 - \ldots - \sigma_j^2) \leq 0 \quad [3.33]$$

Because $\eta_i \in \mathbf{B}^{n_y}$, with $i = 1, \ldots, j$, the following expression is obtained:

$$\begin{bmatrix} \sigma_1^2 & \ldots & \sigma_j^2 \end{bmatrix} \left(\begin{bmatrix} 1 \\ \vdots \\ 1 \end{bmatrix} - \begin{bmatrix} \eta_1^2 \\ \vdots \\ \eta_j^2 \end{bmatrix} \right) \geq 0 \quad [3.34]$$

Adding this term to the left side of [3.33] leads to the following sufficient condition for [3.33]:

$$\hat{\mathbf{z}}^T \hat{H}_{k/j}^T P \hat{H}_{k/j} \hat{\mathbf{z}} - \beta \mathbf{z}^T \hat{H}_{k-1}^T P \hat{H}_{k-1} \mathbf{z} - \mathbf{s}^T F^T F \mathbf{s} - \sigma_1^2 \eta_1^2 - \ldots - \sigma_j^2 \eta_j^2 \leq 0 \quad [3.35]$$

Similar to the Single-Output case, expression [3.35] is equivalent to:

$$\begin{bmatrix}\boldsymbol{\theta}\\ \mathbf{s}\\ \eta_1\\ \vdots\\ \eta_j\end{bmatrix}^T \begin{bmatrix}\beta P & 0 & 0 & \dots & 0 & B_1\\ * & F^TF & 0 & \dots & 0 & B_2\\ * & * & \sigma_1^2 & \dots & 0 & B_3\\ \vdots & \vdots & \vdots & \ddots & \vdots & \vdots\\ * & * & * & \dots & \sigma_j^2 & B_{j+2}\\ * & * & * & \dots & * & P\end{bmatrix} \begin{bmatrix}\boldsymbol{\theta}\\ \mathbf{s}\\ \eta_1\\ \vdots\\ \eta_j\end{bmatrix} > 0, \forall \boldsymbol{\theta}, \mathbf{s}, \eta_1, \dots, \eta_j \quad [3.36]$$

with $\boldsymbol{\theta} = H_{k/j}\mathbf{z}$ and the notations:

$$\begin{aligned}
B_1 &= ((\prod_{i=1}^{j}(I - \boldsymbol{\lambda}_{j+1-i}\mathbf{c}_{j+1-i}^T))A)^T P\\
B_2 &= ((\prod_{i=1}^{j}(I - \boldsymbol{\lambda}_{j+1-i}\mathbf{c}_{j+1-i}^T))F)^T P\\
B_3 &= (\prod_{i=1}^{j-1}(I - \boldsymbol{\lambda}_{j+1-i}\mathbf{c}_{j+1-i}^T)\sigma_1\boldsymbol{\lambda}_1)^T P\\
&\vdots\\
B_j &= ((I - \boldsymbol{\lambda}_j\mathbf{c}_j^T)(I - \boldsymbol{\lambda}_{j-1}\mathbf{c}_{j-1}^T)\sigma_{j-2}\boldsymbol{\lambda}_{j-2})^T P\\
B_{j+1} &= ((I - \boldsymbol{\lambda}_j\mathbf{c}_j^T)\sigma_{j-1}\boldsymbol{\lambda}_{j-1})^T P\\
B_{j+2} &= (\sigma_j\boldsymbol{\lambda}_j)^T P
\end{aligned} \quad [3.37]$$

Using the definition of positive definite matrix (see Appendix 1) and the minimization of the P-radius, the following optimization problem is obtained:

$$\max_{\tau,\beta,P,Y_j} \tau$$

subject to

$$\begin{cases} \dfrac{(1-\beta)P}{\sigma_1^2+\ldots+\sigma_j^2+\max\limits_{\mathbf{s}\in\mathbf{B}^{n_x}}\|F\mathbf{s}\|_2^2} \succeq \tau I \\ \begin{bmatrix} \beta P & 0 & 0 & \ldots & 0 & B_1 \\ * & F^TF & 0 & \ldots & 0 & B_2 \\ * & * & \sigma_1^2 & \ldots & 0 & B_3 \\ \vdots & \vdots & \vdots & \ddots & \vdots & \vdots \\ * & * & * & \ldots & \sigma_j^2 & B_{j+2} \\ * & * & * & \ldots & * & P \end{bmatrix} \succeq 0 \\ \tau > 0 \end{cases} \qquad [3.38]$$

As the vectors $\boldsymbol{\lambda}_1$, ..., $\boldsymbol{\lambda}_{j-1}$ are computed in the previous steps, [3.38] is a BMI optimization problem whose decision variables are $Y_j = P\boldsymbol{\lambda}_j$, $P = P^T$, β and τ.

EXAMPLE 3.2.– Consider the following uncertain two-output linear discrete-time invariant system:

$$\begin{cases} \mathbf{x}_{k+1} = \begin{bmatrix} 0 & -0.5 \\ 1 & 1+0.3\delta \end{bmatrix}\mathbf{x}_k + 0.02\begin{bmatrix} -6 \\ 1 \end{bmatrix}\omega_k \\ \mathbf{y}_k = \begin{bmatrix} -2 & 1 \\ 1 & 1 \end{bmatrix}\mathbf{x}_k + \begin{bmatrix} 0.2 & 0 \\ 0 & 0.2 \end{bmatrix}\mathbf{v}_k \end{cases} \qquad [3.39]$$

with $\|\nu_k\|_\infty \leq 1$, $\|\omega_k\|_\infty \leq 1$ and $|\delta| \leq 1$.

In this example, the results obtained by the ESO approach (Algorithm 3.1) and ESOCE approach (Algorithm 3.2) are compared with the results obtained by the segment minimization approach and the volume minimization approach [ALA 05] applied for the multivariable case. The ESO approach gives the correction factors $\boldsymbol{\lambda}_1 = \begin{bmatrix} -0.2137 \\ 0.5726 \end{bmatrix}$ and $\boldsymbol{\lambda}_2 = \begin{bmatrix} 0.3684 \\ 0.3570 \end{bmatrix}$. The correction factors computed by Algorithm 3.2 (ESOCE) are $\boldsymbol{\lambda}_1 = \begin{bmatrix} -0.2137 \\ 0.5726 \end{bmatrix}$ and $\boldsymbol{\lambda}_2 = \begin{bmatrix} 0.2839 \\ 0.5085 \end{bmatrix}$.

The simulation result shows good accuracy of the proposed approaches. The volume minimization approach gives the best result of the estimation with an important online computation time. The segment minimization approach offers a fast computation with a degradation of the guaranteed state estimation set. The proposed approaches provide good trade-off solutions between the complexity of the computation and the accuracy of the state estimation (Figures 3.13–3.15; Table 3.2).

Figure 3.11 shows the decrease of the zonotopic guaranteed state estimation at each time instant ($k = 1$, $k = 2$, $k = 3$). Figure 3.12 shows the guaranteed bounds on x_1 obtained with different approaches and the real states (which are in between the guaranteed bounds). Figures 3.13–3.15 show that the accuracy of the ESOCE algorithm is better than the accuracy of the ESO algorithm which confirms a less conservative result of the ESOCE method. In addition, the results obtained using the ESOCE algorithm are closer (than the results obtained by the ESO approach) to the results obtained by the volume minimization of the zonotope.

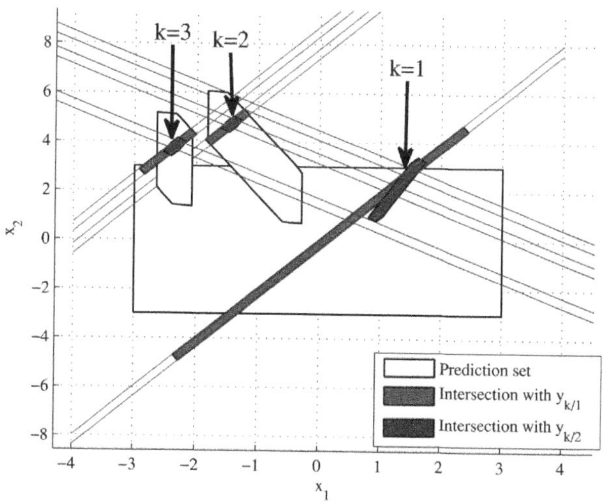

Figure 3.11. *Intersection $\hat{\mathcal{X}}_k$ between the predicted state set $\bar{\mathcal{X}}_k$ and the measurement $\mathcal{X}_{y_k}$ using ESOCE approach (Example 3.2). For a color version of this figure, see www.iste.co.uk/stoica/zonotopes.zip*

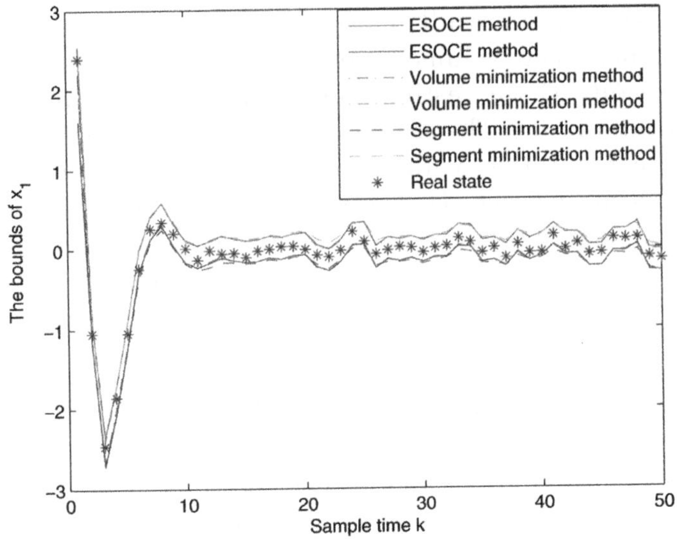

Figure 3.12. *Guaranteed bound of x_1 via different methods (Example 3.2)*

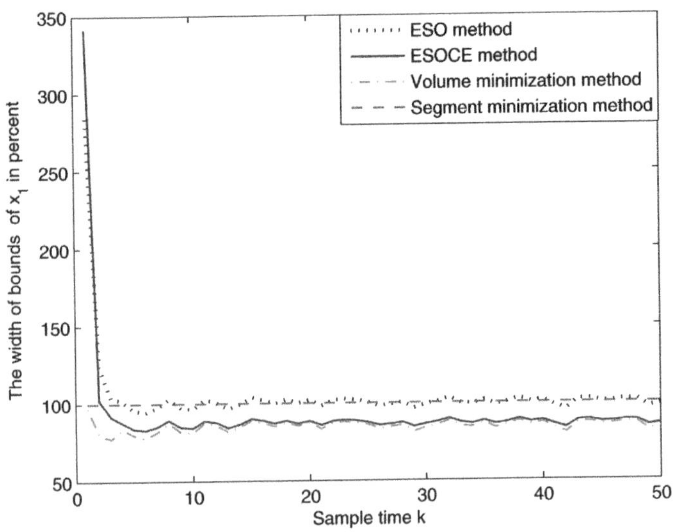

Figure 3.13. *Comparison of the bound's width of x_1 via different methods in percent (Example 3.2)*

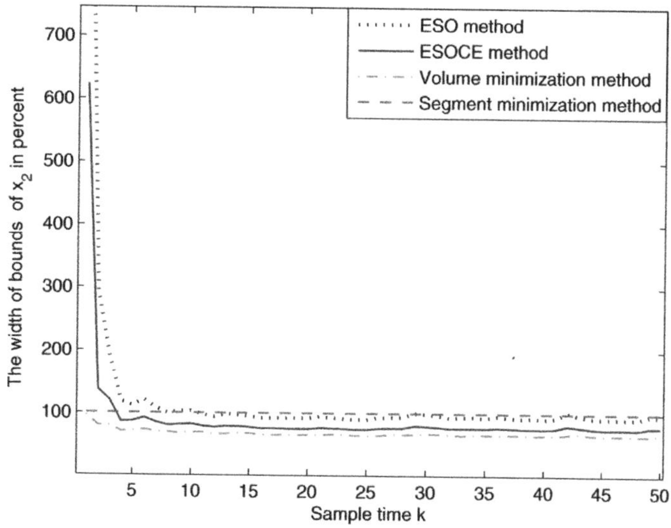

Figure 3.14. *Comparison of the bound's width of x_2 via different methods in percent (Example 3.2)*

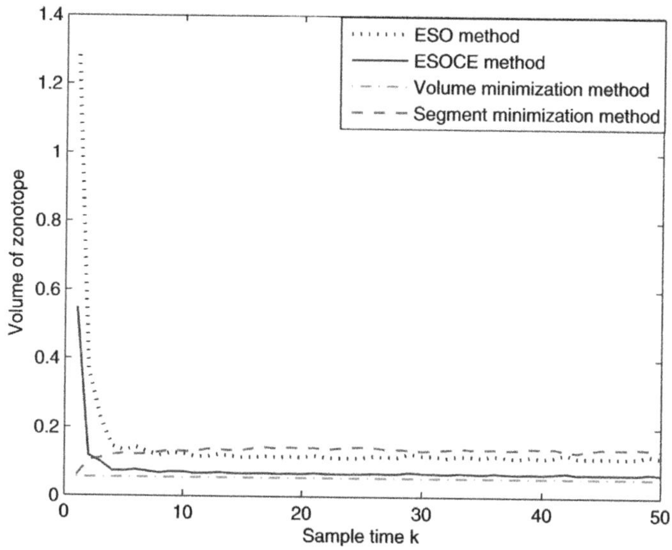

Figure 3.15. *Comparison of the volume of zonotopic state estimation set via different methods (Example 3.2)*

Algorithm	Time (seconds)
Segment minimization	0.0624
Presented algorithms (ESO and ESOCE) (without off-line optimization included)	0.0624
ESO algorithm (with off-line optimization included)	1.3884
ESOCE algorithm (with off-line optimization included)	1.4664
Volume minimization	27.2534

Table 3.2. *Total computation time of Example 3.2 after 50 time instants*

Figures 3.16 and 3.17 compare the accuracy of the estimation when the order of taking into account the measurements is changed (the vector $\mathbf{c}_1$ is replaced by $\mathbf{c}_2$ and vice versa). The comparison leads to the following conclusion: the order of taking into account the measurement can influence the accuracy of the estimation. Thus, to obtain the best result of the approximation in the case of Multi-Output systems, $n_y!$ different combinations of output measurements must be tried and compared, and then the best solution must be chosen for the implementation. The Singular Value Decomposition (SVD) method, which is a method dedicated to Multi-Output systems, has the same problem (expression [2.12] is different when the rows of the C matrix are changed and, thus, the accuracy of the estimation is influenced). Note that for systems with a large number of outputs, comparison of the results of all the possible combinations can be time consuming. To overcome the conservativeness of these approaches, a new method which considers a unique coupled Single-Output-based optimization problem is proposed in the following section.

3.2.2.3. *Unique coupled Single-Output-based approach*

In the previous approaches, the vectors $\boldsymbol{\lambda}_1, \ldots, \boldsymbol{\lambda}_{n_y}$ are separately computed which can lead to a conservative result. To overcome this problem, the solution proposed now is to compute $\boldsymbol{\lambda}_1, \ldots, \boldsymbol{\lambda}_{n_y}$ at the same time. All the vectors $\boldsymbol{\lambda}_i$, with $i = 1, \ldots, n_y$, are computed such that the condition on the non-increasing property of the P-radius is ensured. This condition leads to a similar inequality as [3.33], with the modification that j is replaced by n_y.

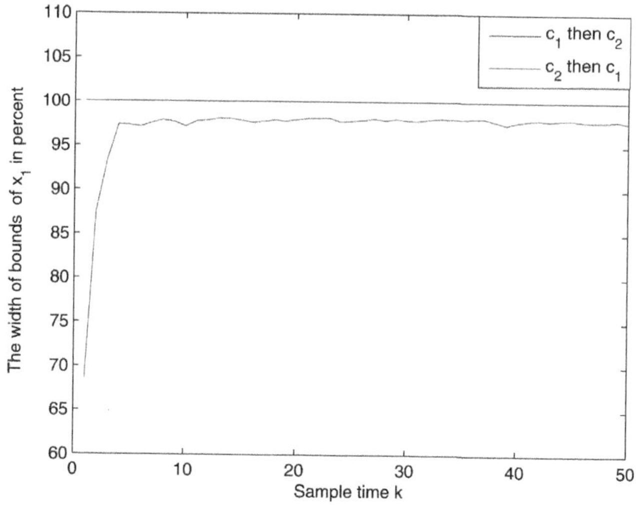

Figure 3.16. *Comparison of the bound's width of x_1 with interchanged output measurements (Example 3.2). For a color version of this figure, see www.iste.co.uk / stoica / zonotopes.zip*

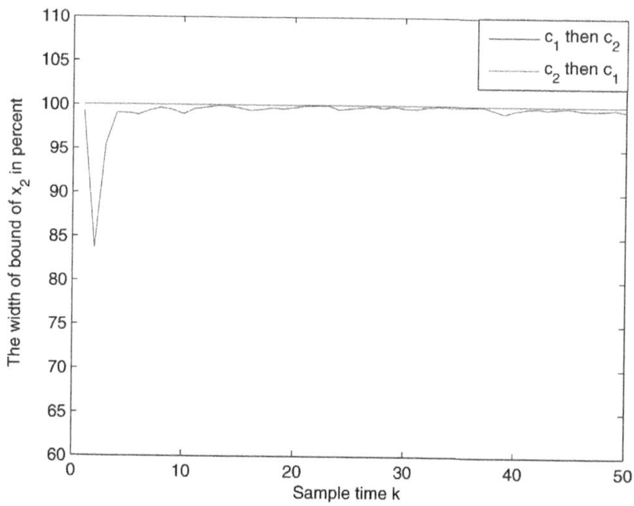

Figure 3.17. *Comparison of the bound's width of x_2 with interchanged output measurements (Example 3.2). For a color version of this figure, see www.iste.co.uk / stoica / zonotopes.zip*

Similar to the ESOCE approach, the following optimization problem is obtained:

$$\max_{\tau,\beta,P,\boldsymbol{\lambda}_1,\ldots,\boldsymbol{\lambda}_{ny}} \quad \tau$$

subject to

$$\begin{cases} \frac{(1-\beta)P}{\sigma_1^2+\ldots+\sigma_{ny}^2+\max\limits_{\mathbf{s}\in\mathbf{B}^{nx}}\|F\mathbf{s}\|_2^2} \succeq \tau I \\ \begin{bmatrix} \beta P & 0 & 0 & \ldots & 0 & B_1 \\ * & F^TF & 0 & \ldots & 0 & B_2 \\ * & * & \sigma_1^2 & \ldots & 0 & B_3 \\ \vdots & \vdots & \vdots & \ddots & \vdots & \vdots \\ * & * & * & \ldots & \sigma_{ny}^2 & B_{n_y+2} \\ * & * & * & \ldots & * & P \end{bmatrix} \succeq 0 \\ \tau > 0 \end{cases} \quad [3.40]$$

with the notations:

$$\begin{aligned} B_1 &= ((\prod_{i=1}^{n_y}(I-\boldsymbol{\lambda}_{n_y+1-i}\mathbf{c}_{n_y+1-i}^T))A)^TP \\ B_2 &= ((\prod_{i=1}^{n_y}(I-\boldsymbol{\lambda}_{n_y+1-i}\mathbf{c}_{n_y+1-i}^T))F)^TP \\ B_3 &= (\prod_{i=1}^{n_y-1}(I-\boldsymbol{\lambda}_{n_y+1-i}\mathbf{c}_{n_y+1-i}^T)\sigma_1\boldsymbol{\lambda}_1)^TP \\ &\vdots \\ B_{n_y} &= ((I-\boldsymbol{\lambda}_{n_y}\mathbf{c}_{n_y}^T)(I-\boldsymbol{\lambda}_{n_y-1}\mathbf{c}_{n_y-1}^T)\sigma_{n_y-2}\boldsymbol{\lambda}_{n_y-2})^TP \\ B_{n_y+1} &= ((I-\boldsymbol{\lambda}_{n_y}\mathbf{c}_{n_y}^T)\sigma_{n_y-1}\boldsymbol{\lambda}_{n_y-1})^TP \\ B_{n_y+2} &= (\sigma_{n_y}\boldsymbol{\lambda}_{n_y})^TP \end{aligned} \quad [3.41]$$

As the vectors $\boldsymbol{\lambda}_1$, …, $\boldsymbol{\lambda}_{n_y}$ are decision variables, the optimization problem [3.40] is no longer a BMI optimization problem as discussed in the previous section. From the definition of Polynomial Matrix Inequalities (PMIs – see Appendix), problem [3.40] is a PMI optimization problem. PMI problems can be solved using several relaxation techniques [CHE 00, HEN 02, KOJ 03, HOL 04, HEN 06]. In this book, the method presented in [HEN 06] is chosen to solve the PMI optimization problem [3.40]. This method is based on an insertion of lifting variables to reduce the polynomial terms to linear terms. Using these new variables the PMI problem becomes a LMI problem, which can be easily solved (see [HEN 06] for more details).

In PMI problem [3.40], with the notations in [3.41], the decision variables are $P = P^T \in \mathbb{R}^{n_x \times n_x}$, $\boldsymbol{\lambda}_1, \ldots, \boldsymbol{\lambda}_{n_y} \in \mathbb{R}^{n_x}$ and the scalar variables β and τ. Thus, the total number of scalar decision variables is $\frac{n_x(n_x+1)}{2} + n_y n_x + 2$. The degree of PMI [3.40] is $n_y + 1$. This PMI can be solved using the first-order LMI relaxation methodology [HEN 06]. In order to understand this better the first-order LMI relaxation method [HEN 06] is illustrated in the following example.

EXAMPLE 3.3.– Consider the following optimization problem:

$$\min_x (x_1^2 - x_2^2)$$

subject to PMI: $\begin{bmatrix} 1 + x_1 x_2 & x_1 \\ x_1 & 1 - x_1^2 - x_2^2 \end{bmatrix} \succeq 0$

First, the following change of variables is made: $y_{10} = x_1$, $y_{01} = x_2$, $y_{20} = x_1^2$, $y_{02} = x_2^2$, $y_{11} = x_1 x_2$.

Using these new variables, the PMI optimization problem is relaxed by the following LMIs:

$$\min_x (y_{20} - y_{02})$$

subject to LMIs:

$$\left\{ \begin{array}{l} \begin{bmatrix} 1 & y_{10} & y_{01} \\ * & y_{20} & y_{11} \\ * & * & y_{02} \end{bmatrix} \succeq 0 \\ \begin{bmatrix} 1 + y_{11} & y_{10} \\ * & 1 - y_{20} - y_{02} \end{bmatrix} \succeq 0 \end{array} \right. \qquad [3.42]$$

Using this technique, the PMI optimization [3.40] can be solved as in the following procedure [HEN 06].

First, all the $\frac{n_x(n_x+1)}{2} + n_y n_x + 2$ *scalar decision variables* are denoted as follows: τ, $y_{10\ldots0}$, $y_{01\ldots0}$, $\ldots$, $y_{00\ldots10}$, $y_{00\ldots01}$. For expression [3.40], this leads to $\beta = y_{100\ldots0}$, $P = \begin{bmatrix} y_{01\ldots0} \cdots y_{00\ldots1\ldots0} \\ \cdots \quad \cdots \quad \cdots \end{bmatrix}$, $\boldsymbol{\lambda}_1^T = \begin{bmatrix} y_{00\ldots1\ldots0} \cdots \end{bmatrix}$, $\ldots$, $\boldsymbol{\lambda}_{n_y}^T = \begin{bmatrix} y_{00\ldots1\ldots0} \cdots y_{00\ldots01} \end{bmatrix}$.

Second, the *polynomial decision elements* in [3.40] are rewritten as the result of a change of variables based on the previous scalar decision variables $y_{20\ldots0} = y_{10\ldots0} \cdot y_{10\ldots0}$, $y_{11\ldots0} = y_{10\ldots0} \cdot y_{01\ldots0}$, etc. In this way, expression [3.40] becomes an LMI. If the LMI relaxation of the PMI optimization problem [3.40] is used, then the following LMI problem must be solved (see [HEN 06] for more details):

$$\max \tau$$

subject to the LMIs

$$\left\{ \begin{array}{l} M_1 = \begin{bmatrix} 1 & * & * & * \\ y_{10\ldots0} & y_{20\ldots0} & \cdots & * \\ y_{01\ldots0} & y_{11\ldots0} & \cdots & * \\ \vdots & \vdots & \ddots & \vdots \\ y_{00..111} & \cdots & \cdots & y_{00\ldots222} \end{bmatrix} \succeq 0 \\ \\ M_2 \succeq 0 \end{array} \right. \qquad [3.43]$$

where M_1 is called the moment matrix and M_2 denotes the equivalent LMI expressions obtained from the PMI in [3.40] using the new scalar decision variables ($y_{10\ldots0}$, $y_{20\ldots0}$, etc.).

As the degree of the PMI in [3.40] is $n_y + 1$, the dimensions of LMIs [3.43] are:

– M_1: $q \times q$, with $q = 1 + \frac{n_x(n_x+1)}{2} + n_y n_x + 1 + \left(\frac{n_x(n_x+1)}{2} + n_y n_x + 1\right)^2 + \ldots + \left(\frac{n_x(n_x+1)}{2} + n_y n_x + 1\right)^{round(\frac{n_y+1}{2})}$;

– M_2: $n_x \times n_x$, $(3n_x + n_y) \times (3n_x + n_y)$, and 1.

The number of scalar decision variables in this optimization problem is $\frac{q(q+1)}{2}$.

Using the relaxation method proposed in [HEN 06] allows us to relax the PMI optimization problem to a LMI optimization problem which is easier to solve. Even if the obtained LMIs problem is solved off-line, this method leads to a large LMI problem (see the size of matrix M_1 and the number of scalar decision variables) which limits its application.

EXAMPLE 3.4.– Consider the following two-output linear discrete-time invariant system:

$$\begin{cases} \mathbf{x}_{k+1} = \begin{bmatrix} 0 & -0.5 \\ 1 & 1 \end{bmatrix} \mathbf{x}_k + 0.02 \begin{bmatrix} -6 \\ 1 \end{bmatrix} \omega_k \\ \mathbf{y}_k = \begin{bmatrix} -2 & 1 \\ 1 & 1 \end{bmatrix} x_k + 0.2\mathbf{v}_k \end{cases} \qquad [3.44]$$

with $\|\mathbf{v}_k\|_\infty \leq 1$ and $\|\omega_k\|_\infty \leq 1$. The values of $\mathbf{v}_k$ and ω_k are generated by the random functions of $Matlab^{®}$. The initial state is unknown and belongs to the box $3\mathbf{B}^2$.

Denote $\beta = y_{10000000}$, $P = \begin{bmatrix} y_{01000000} & y_{00100000} \\ * & y_{00010000} \end{bmatrix}$, $\boldsymbol{\lambda}_1^T = \begin{bmatrix} y_{00001000} & y_{00000100} \end{bmatrix}$, $\boldsymbol{\lambda}_2^T = \begin{bmatrix} y_{00000010} & y_{00000001} \end{bmatrix}$. Thus, the dimensions of the LMIs of the considered optimization problem [3.43] are:

– M_1: 37×37;

– M_2: 2×2, 7×7, and 1.

As $l = 8$ and $q = 37$, the total number of scalar decision variables of the LMI problem [3.43] is $\frac{37 \cdot 38}{2} + 1 = 704$, which is still reasonable for any LMI solver.

Figures 3.18 and 3.19 show the evolution of the predicted state set and the outer approximation of the state estimation set at time instant $k = 1$ and $k = 8$, respectively. Comparing these figures confirms the contractiveness of the guaranteed state estimation.

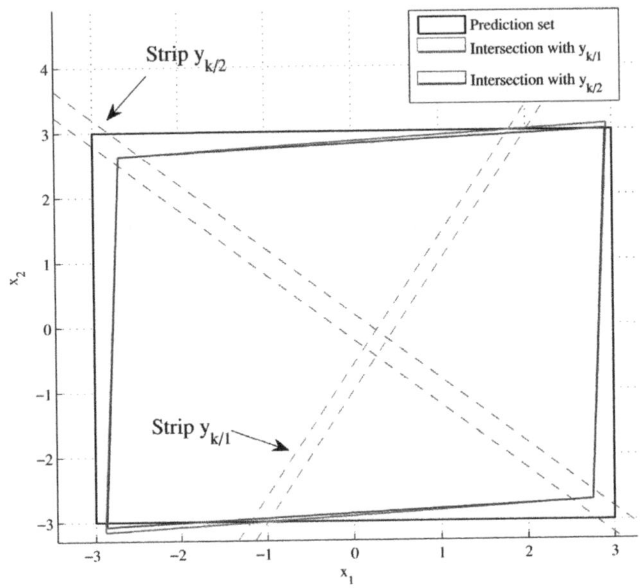

Figure 3.18. *Guaranteed state estimation at time instant $k = 1$ (Example 3.4). For a color version of this figure, see www.iste.co.uk/stoica/zonotopes.zip*

Figures 3.20 and 3.21 show the obtained guaranteed bounds on x_1 and x_2, respectively. The real states (star) are found inside these bounds, which confirms good accuracy of the estimation. Figure 3.22 shows the evolution of the volume of the zonotopic guaranteed state estimation. Note that these guaranteed bounds and the volume of zonotope are decreased in time leading to more and more accurate estimation.

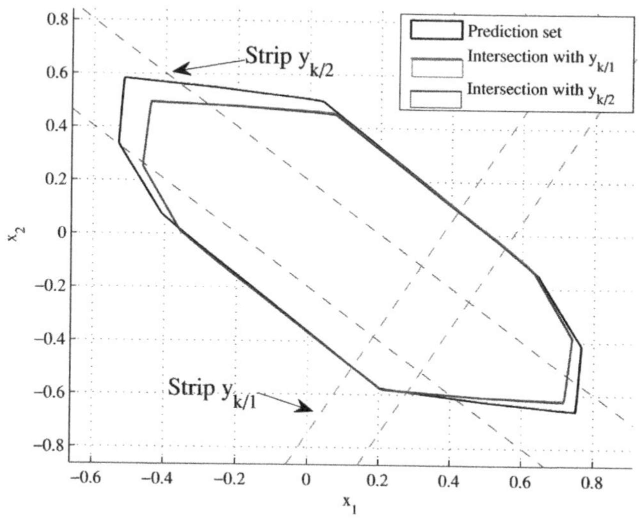

Figure 3.19. *Guaranteed state estimation at time instant $k = 8$ (Example 3.4). For a color version of this figure, see www.iste.co.uk / stoica / zonotopes.zip*

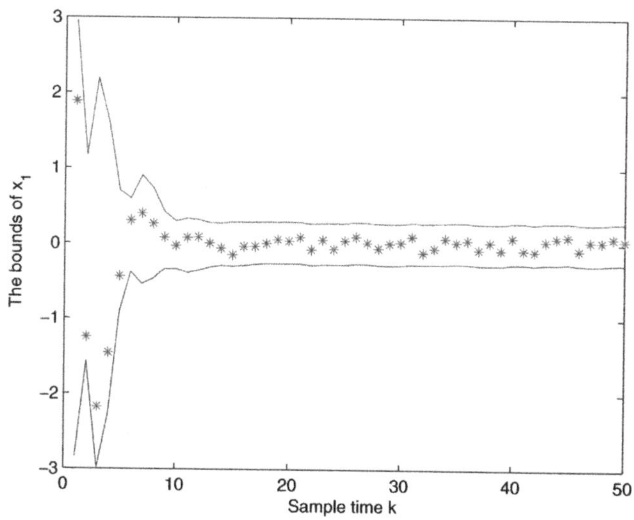

Figure 3.20. *Guaranteed bounds of x_1 by the PMI-based approach (Example 3.4). For a color version of this figure, see www.iste.co.uk / stoica / zonotopes.zip*

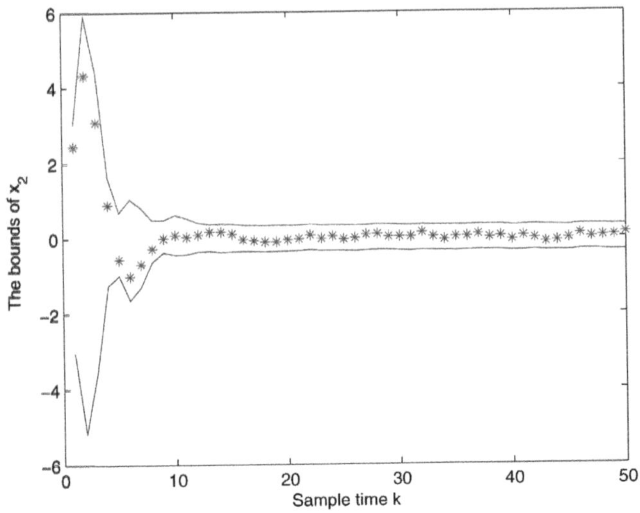

Figure 3.21. *Guaranteed bounds of x_2 by the PMI-based approach (Example 3.4). For a color version of this figure, see www.iste.co.uk / stoica / zonotopes.zip*

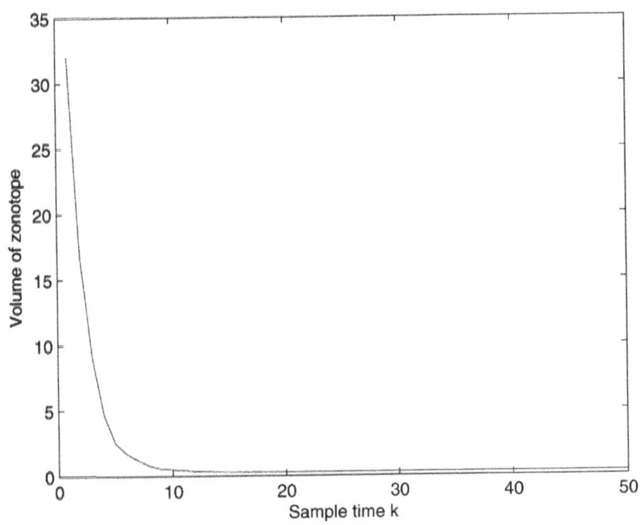

Figure 3.22. *Evolution of the volume of the guaranteed state estimation obtained by the PMI-based approach (Example 3.4)*

Table 3.3 shows the computation time of the PMI approach after 50 time instants. Even if the solution of the PMI problem is computed off-line, the computation time of this approach is significantly increased due to the large number of decision variables in the LMIs problem [3.43]. Even if the contractiveness property of the guaranteed state estimation is always preserved, the proposed method gives an unsatisfactory result of estimation. The poor result of estimation obtained in this example can be due to the relaxation method used to solve the PMI optimization problem. To reduce the conservatism of this solution, a higher order of the LMI relaxation (a higher order of the moment matrix M_1) can be used leading to a more complex LMI optimization problem than the first-order LMI relaxation (see [HEN 06]). In the recent report [HEN 11], the author proposed an inner approximation of the PMI set, instead of the outer approximation [HEN 06] used in this example, which can lead to a less conservative result of the guaranteed state estimation.

Algorithm	Time (seconds)
PMI algorithm (without off-line LMI-relaxed optimization included)	0.0780
PMI algorithm (with off-line LMI-relaxed optimization included)	3436.1

Table 3.3. *Total computation time of Example 3.4 after 50 time instants*

To conclude, Table 3.4 recapitulates the complexity and the accuracy of the proposed approaches to solve the estimation problem for the Multi-Output system, numbers from 1 to 5 being associated with each approach. The number 1 means the best accuracy and the less complexity of the computation. The number 5 means the worst accuracy and the most complex from the computation point of view.

To avoid solving a PMI problem which can lead to an unaccurate result (due to the choice of the relaxation procedure), in the following section a solution based on the intersection of a polytope and a zonotope will be proposed.

Algorithm	Complexity	Accuracy
Segment minimization method	1	4
Volume minimization method	4	1
ESO method	2	3
ESOCE method	2	2
PMI method	5	5

Table 3.4. *Table of recapitulation of different estimation approaches (Multi-Output case)*

3.2.3. *Dedicated approach for Multi-Output systems*

In the previous section 3.2.2, the guaranteed state estimation set is obtained using step-by-step algorithms (the ESO method, ESOCE method and PMI-based method). This means that the predicted state set intersects with the first measurement leading to a zonotopic outer approximation of this intersection. Then this approximation is intersected with the second measurement leading to a zonotopic outer approximation of this intersection and so on. This step-by-step algorithm does not really respect Algorithm 2.2, where the measurement consistent state set induced by the measurement is intersected with the predicted state set. Moreover, the order used to take into account the different measurements is important. The different order used gives a different result of the state estimation, thus all the possible rearrangements of the output vector components must be tested to find the optimum order. All these tests are increasing the computation complexity.

The solution, called Polytope and Zonotope Intersection (PAZI) [LE 12b, LE 13], proposed in this section consists of directly intersecting the measurement consistent state set with the predicted state set. As each measurement is represented by a strip $|\mathbf{c}_i^T\mathbf{x} - y_{k/i}| \leq \sigma_i$, the measurement consistent state set is the intersection of n_y strips or more precisely a H-polytope. As the predicted state set is a zonotope, it is convenient to determine the intersection between a zonotope and a polytope.

The problem of finding the intersection between a zonotope $\mathcal{Z}$ and a polytope $\mathcal{P}$ can be solved in an easy way: the zonotope $\mathcal{Z}$ is transformed from the generator representation to the H-representation. Then the intersection between $\mathcal{Z}$ and $\mathcal{P}$ can be easily computed using the Multi-Parametric Toolbox (MPT) [KVA 04]. This way gives an exact computation of the intersection between the zonotope $\mathcal{Z}$ and the polytope $\mathcal{P}$. Unfortunately, this intersection is not always a zonotope and is represented by a polytope in the H-representation. In order to continue the next step, an outer approximation of this intersection (i.e. polytope) by a zonotope must be found. In the literature, the approximation of a polytope by a zonotope is studied by many authors; however, this problem still remains complex (see [STU 03, BAR 99, GUI 05]). Therefore, to overcome this difficult problem, a simpler way to obtain a zonotope in the generator representation, characterizing the intersection between a polytope and a zonotope, is presented below.

PROPOSITION 3.2.– [LE 13].– Given the zonotope $\mathcal{Z} = \mathbf{p} \oplus H\mathbf{B}^r \subset \mathbb{R}^n$, the polytope $\mathcal{P} = \{\mathbf{x} \in \mathbb{R}^n : |C\mathbf{x} - \mathbf{d}| \leq \begin{bmatrix} \sigma_1 \\ \vdots \\ \sigma_m \end{bmatrix}\}$ ($\mathbf{d} \in \mathbb{R}^m$, $\sigma_i \in \mathbb{R}^+$, $i = 1, \ldots, m$) and the matrix $\Lambda \in \mathbb{R}^{n \times m}$, define a vector $\hat{\mathbf{p}}(\Lambda) = \mathbf{p} + \Lambda(\mathbf{d} - C\mathbf{p}) \in \mathbb{R}^{n_x}$ and a matrix $\hat{H}(\Lambda) = \begin{bmatrix}(I - \Lambda C)H \ \Lambda\Sigma\end{bmatrix}$, with $\Sigma = \begin{bmatrix} \sigma_1 & 0 & \ldots & 0 \\ 0 & \sigma_2 & \ldots & 0 \\ \vdots & \vdots & \ddots & \vdots \\ 0 & 0 & \ldots & \sigma_m \end{bmatrix} \in \mathbb{R}^{m \times m}$ a diagonal matrix. Then a family of zonotopes (parameterized by the matrix Λ) that contains the intersection of the zonotope $\mathcal{Z}$ and the polytope $\mathcal{P}$ is obtained such that $\mathcal{Z} \cap \mathcal{P} \subseteq \hat{\mathcal{Z}}(\Lambda) = \hat{\mathbf{p}}(\Lambda) \oplus \hat{H}(\Lambda)B^{r+m}$.

PROOF.– Consider an element $\mathbf{x} \in \mathcal{Z} \cap \mathcal{P}$; on the one hand, this means that $\mathbf{x} \in \mathcal{Z} = \mathbf{p} \oplus H\mathbf{B}^r$. Using the definition of a zonotope implies that there exists a vector $\mathbf{s}_1 \in \mathbf{B}^r$ such that:

$$\mathbf{x} = \mathbf{p} + H\mathbf{s}_1 \qquad [3.45]$$

Adding and subtracting $\Lambda CH\mathbf{s}_1$ to the previous equality lead to the following expression:

$$\mathbf{x} = \mathbf{p} + \Lambda CH\mathbf{s}_1 + (I - \Lambda C)H\mathbf{s}_1 \qquad [3.46]$$

On the other hand, from $\mathbf{x} \in \mathcal{Z} \cap \mathcal{P}$, it is inferred that:

$$\mathbf{x} \in \mathcal{P} = \{x \in \mathbb{R}^n : |C\mathbf{x} - \mathbf{d}| \leq \begin{bmatrix} \sigma_1 \\ \vdots \\ \sigma_m \end{bmatrix}\} \qquad [3.47]$$

Thus, a vector $\mathbf{s}_2 \in \mathbf{B}^m$ exists such that $C\mathbf{x} - \mathbf{d} = \Sigma\mathbf{s}_2$. Taking into account the form of the vector $\mathbf{x}$ given by [3.45] leads to $C(\mathbf{p} + H\mathbf{s}_1) - \mathbf{d} = \Sigma\mathbf{s}_2$, which is equivalent to $CH\mathbf{s}_1 = \mathbf{d} - C\mathbf{p} + \Sigma\mathbf{s}_2$. Substituting $CH\mathbf{s}_1$ in equation [3.46], the following expression is obtained:

$$\begin{aligned} \mathbf{x} &= \mathbf{p} + \Lambda(\mathbf{d} - C\mathbf{p} + \Sigma\mathbf{s}_2) + (I - \Lambda C)H\mathbf{s}_1 \\ &= \mathbf{p} + \Lambda(\mathbf{d} - C\mathbf{p}) + \Lambda\Sigma\mathbf{s}_2 + (I - \Lambda C)H\mathbf{s}_1 \end{aligned} \qquad [3.48]$$

After simple computations and using the notation defined in Proposition 3.2, the following form is obtained:

$$\mathbf{x} = \hat{\mathbf{p}}(\Lambda) + \left[(I - \Lambda C)H \; \Lambda\Sigma\right] \begin{bmatrix} \mathbf{s}_1 \\ \mathbf{s}_2 \end{bmatrix} = \hat{\mathbf{p}}(\Lambda) \oplus \hat{H}(\Lambda) \begin{bmatrix} \mathbf{s}_1 \\ \mathbf{s}_2 \end{bmatrix} \qquad [3.49]$$

and the following inclusion holds:

$$\mathbf{x} = \hat{\mathbf{p}}(\Lambda) \oplus \hat{H}(\Lambda) \begin{bmatrix} \mathbf{s}_1 \\ \mathbf{s}_2 \end{bmatrix} \in \hat{\mathbf{p}}(\Lambda) \oplus \hat{H}(\Lambda)\mathbf{B}^{r+m} = \hat{\mathcal{X}}(\Lambda). \quad [3.50]$$

□

This proposition is illustrated by the following example.

EXAMPLE 3.5.– Consider a centered zonotope $\mathcal{Z}(\mathbf{p}; H)$, with $\mathbf{p} = \begin{bmatrix} 0 \; 0 \end{bmatrix}$ and $H = \begin{bmatrix} 0.1 \; 0.2 \; 0.3 \\ 0.3 \; 0.2 \; 0.1 \end{bmatrix}$, and a polytope $\mathcal{P}$ described by the intersection of three strips:

$$|C\mathbf{x} - \mathbf{d}| < \begin{bmatrix} \sigma_1 \\ \sigma_2 \\ \sigma_3 \end{bmatrix}, \text{ with } C = \begin{bmatrix} 5 & 1 \\ -4 & 1 \\ 1 & 2 \end{bmatrix}, \mathbf{d} = \begin{bmatrix} -0.1163 \\ -0.2935 \\ -0.6928 \end{bmatrix}, \sigma_1 = 0.2,$$

$\sigma_2 = 0.2$, $\sigma_3 = 0.3$.

Figure 3.23 shows a zonotopic outer approximation of the zonotope $\mathcal{Z}$ and the polytope $\mathcal{P}$ using Proposition 3.2. The matrix $\Lambda = \begin{bmatrix} 0.1123 & -0.1044 & 0.0035 \\ 0.1914 & 0.2646 & 0.2026 \end{bmatrix}$ is determined using the minimization of segments of the zonotope. As Λ is a free matrix, the method of computation of Λ can influence the quality of the approximation.

Similar to the previous section 3.2.2, the predicted state set is computed by [2.6]. The measurement consistent state set $\mathcal{X}_{\mathbf{y}_k}$ is a polytope described by:

$$\mathcal{X}_{\mathbf{y}_k} = \{\mathbf{x} \in \mathbb{R}^n : |C\mathbf{x} - \mathbf{y}_k| \leq \begin{bmatrix} \sigma_1 \\ \vdots \\ \sigma_{n_y} \end{bmatrix}\} \quad [3.51]$$

Thus, the exact uncertain state set is the intersection between $\bar{\mathcal{X}}_k$ (a zonotope) and $\mathcal{X}_{\mathbf{y}_k}$ (a polytope). Using Proposition 3.2, the guaranteed state estimation of the

Multi-Output system [2.5] obtained by intersecting the zonotope $\bar{\mathcal{X}}_k$ and the polytope $\mathcal{X}_{\mathbf{y}_k}$ is a family of zonotopes parameterized by the matrix Λ:

$$\hat{\mathcal{X}}_k(\Lambda) = \hat{\mathbf{p}}_k(\Lambda) \oplus \hat{H}_k(\Lambda)\mathbf{B}^{r+n_x+n_y} \qquad [3.52]$$

with $\hat{\mathbf{p}}_k(\Lambda) = A\hat{\mathbf{p}}_{k-1} + \Lambda(\mathbf{y}_k - CA\hat{\mathbf{p}}_{k-1})$
and $\hat{H}_k(\Lambda) = \big[(I - \Lambda C)A\hat{H}_{k-1}\ (I - \Lambda C)F\ \Lambda\Sigma\big]$.

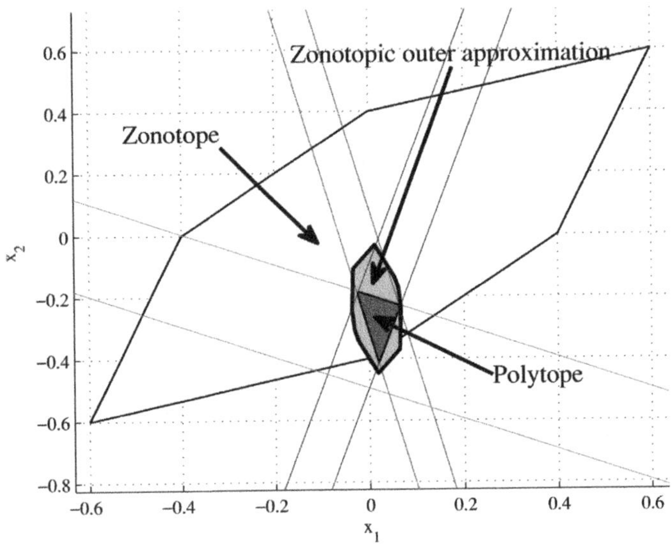

Figure 3.23. *Outer approximation of the intersection between a zonotope and a polytope. For a color version of this figure, see www.iste.co.uk/stoica/zonotopes.zip*

This matrix Λ is computed such that the P-radius of the guaranteed state estimation is non-increasing. Similar to the procedure for the Single-Output case, this condition leads to the following optimization problem:

$$\max_{\tau,\beta,P,Y} \tau$$

subject to

$$\begin{cases} \frac{(1-\beta)P}{\sigma_1^2+\ldots+\sigma_{n_y}^2+\max\limits_{\mathbf{s}\in\mathbf{B}^{n_x}} \|F\mathbf{s}\|_2^2} \succeq \tau I \\ \begin{bmatrix} \beta P & 0 & 0 & A^TP - A^TC^TY^T \\ * & F^TF & 0 & F^TP - F^TC^TY^T \\ * & * & \Sigma^T\Sigma & Y^T\Sigma \\ * & * & * & P \end{bmatrix} \succeq 0 \\ \tau > 0 \end{cases} \qquad [3.53]$$

with a change of variables $Y = P\Lambda$.

In the optimization problem [3.53], the decision variables are $P = P^T \in \mathbb{R}^{n_x \times n_x}$, $Y \in \mathbb{R}^{n_x \times n_y}$, $\beta \in (0, 1]$ and $\tau \in \mathbb{R}$. The total number of scalar decision variables is $\frac{n_x(n_x+1)}{2} + n_x n_y + 2$. The dimensions of matrix inequalities in [3.53] are $n_x \times n_x$, $(3n_x + n_y) \times (3n_x + n_y)$ and 1, respectively. The dimensions of matrix inequalities and the number of scalar decision variables of this method are similar to the other methods (ESO and ESOCE), but the number of matrix inequalities is less than the number of ESO and ESOCE methods.

The goal of computing all of the correction factors λ_i as mentioned in the PMI approach is obtained by the proposed solution in this section. Moreover, the number of scalar decision variables of this optimization problem is significantly decreased in comparison with that of the relaxed LMIs obtained from the PMI approach.

The following schema recaps the proposed methods to solve the estimation problem for Multi-Output systems.

EXAMPLE 3.6.– Consider the following linear discrete-time invariant Multi-Output system:

$$\begin{cases} \mathbf{x}_{k+1} = \begin{bmatrix} 0 & -0.5 \\ 1 & 1+0.3\delta \end{bmatrix} \mathbf{x}_k + \begin{bmatrix} 0.1 \\ 0.1 \end{bmatrix} \omega_k \\ \\ \mathbf{y}_k = \begin{bmatrix} -2 & 1 \\ 1 & 1 \end{bmatrix} \mathbf{x}_k + \begin{bmatrix} 0.2 & 0 \\ 0 & 0.2 \end{bmatrix} \mathbf{v}_k \end{cases} \quad [3.54]$$

The disturbances, the noises and the uncertainties are always considered bounded by ω_k, $\mathbf{v}_k \in \mathbf{B}^2$ and $|\delta| \leq 1$. The initial state is unknown but belongs to the box $3\mathbf{B}^2$. The proposed PAZI method will be compared with the SVD method [COM 03] and the ESOCE method. Note that the comparison is made only with the SVD method because in the open literature only this method can solve the case of Multi-Output systems.

The solution of the PAZI approach [3.53] obtained by the *Penbmi* solver gives $\beta = 0.0452$, $\Lambda = \begin{bmatrix} -0.3136 & 0.2678 \\ 0.3591 & 0.5811 \end{bmatrix}$. Algorithm 3.2 (ESOCE) provides $\lambda_1 = \begin{bmatrix} -0.2108 \\ 0.5784 \end{bmatrix}$, $\lambda_2 = \begin{bmatrix} 0.2946 \\ 0.5420 \end{bmatrix}$.

The simulation results are shown in Figures 3.24–3.26. The guaranteed state estimation obtained by the PAZI method is quickly decreased at each sample time (Figure 3.24). Figures 3.25 and 3.26 compare the accuracy of the PAZI method with that of the SVD method [COM 03] and the ESOCE method. In this example, the bounds obtained by the PAZI method are smaller (compared in percent) than the bounds obtained by the SVD method with a reduced complexity of the computation (see Table 3.5). Moreover, this polytope intersection solution obtains a similar accuracy of the estimation as the accuracy of the ESOCE method but the computation time is significantly decreased (see Table 3.5).

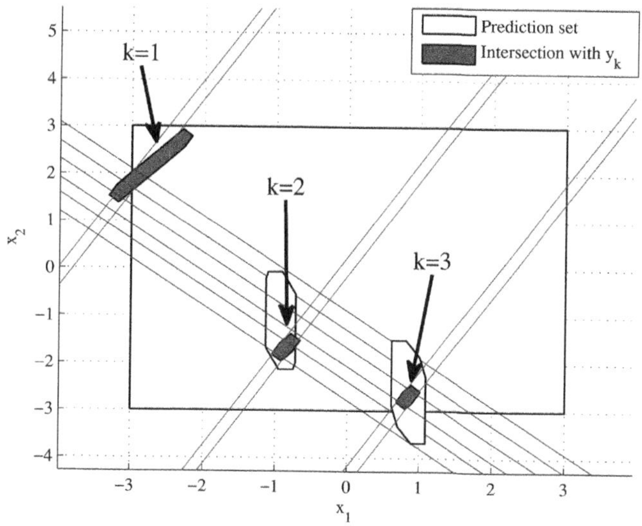

Figure 3.24. *Evolution of the guaranteed state estimation via PAZI approach (Example 3.6). For a color version of this figure, see www.iste.co.uk / stoica / zonotopes.zip*

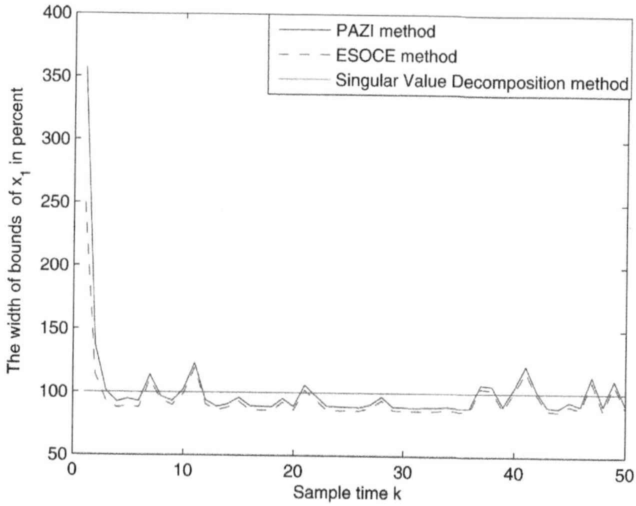

Figure 3.25. *Comparison of the bound's width of x_1 via different methods in percent (Example 3.6). For a color version of this figure, see www.iste.co.uk / stoica / zonotopes.zip*

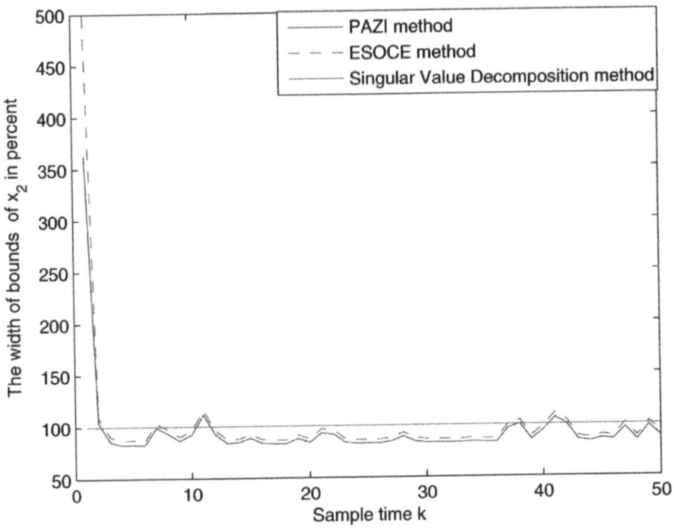

Figure 3.26. *Comparison of the bound's width of x_2 via different methods in percent (Example 3.6). For a color version of this figure, see www.iste.co.uk / stoica / zonotopes.zip*

Algorithm	Time (seconds)
Polytope and zonotope intersection algorithm (without off-line optimization included)	0.0468
Polytope and zonotope intersection algorithm (with off-line optimization included)	0.2808
Singular Value Decomposition algorithm [COM 03]	1.5444

Table 3.5. *Total computation time of Example 4.6 after 50 time instants*

Chapter 4

Tube Model Predictive Control Based on Zonotopic Set-Membership Estimation

This chapter deals with the problem of robust Model Predictive Control (MPC) based on the zonotopic guaranteed set-membership approach for systems with bounded disturbances and measurement noises, under input and state constraints.

4.1. Context

Today, MPC can be considered a mature domain, from both the research and the industrial applications points of view. The use of MPC can be found in many industrial processes (e.g. petroleum industry [GAR 89] and aerospace [BHA 02]), especially when the trajectory to be tracked by the system is given in advance, such as the trajectory of a robot arm [CLA 88]. The large application of MPC [QIN 03] is mainly due to its easy implementation, a generic solution that can deal with both Single-Input Single-Output (SISO) and Multi-Input Multi-Output (MIMO) systems, and, the most importantly, to its ability to handle hard constraints, which often appear when dealing with real plants [CAM 04, GOO 04]. The strategy of MPC can be decomposed into three steps:

1) Based on the appropriate system model, the system output is predicted at a future time horizon.

2) A control sequence is computed by optimizing an objective function on the future behavior of the system, and then only the first element of the control sequence is applied.

3) The horizon is receded into the future, and the computation is repeated at the next time instant.

Because of the receding strategy, the MPC is also called the receding horizon predictive control or moving horizon control. From this algorithm, it must be noted that the role of the system model is important, and thus the choice of the model influences the performance of the MPC strategy. However, if the performance of the controlled system is *not good enough*, the system modeling can be reworked in order to choose a better model [ROS 03].

The receding horizon control strategy is illustrated in Figure 4.1. The future output is predicted over a horizon N_y called the prediction horizon. The control sequence of length N_u ($N_u \leq N_y$) is computed by minimizing an objective function to track the reference trajectory. Then the first element of the control sequence is applied to the system. The difference between the MPC and the classical control is that in the MPC case, the control law is determined from the future error between the system output and the reference, and in the classical control, the action is computed from the error in the past. Thus, when the desired trajectory is known, the MPC technique is more natural than the classical control.

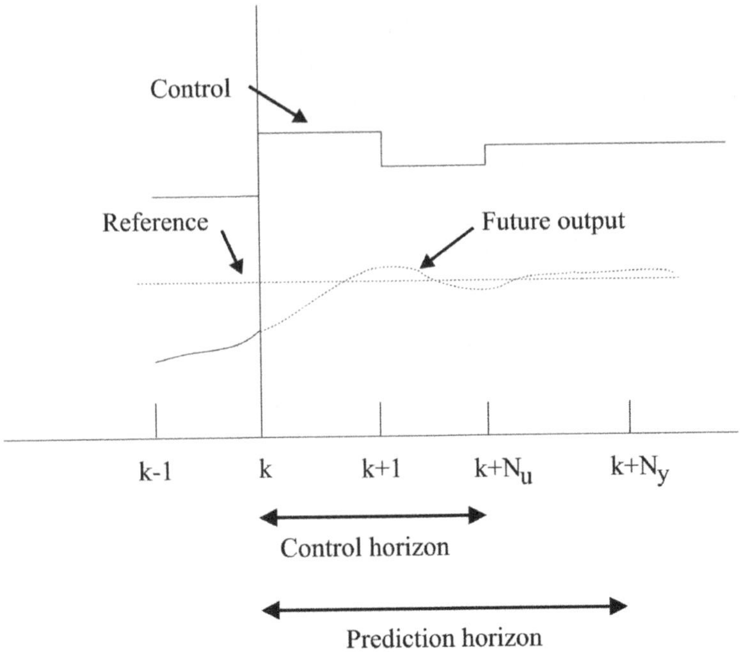

Figure 4.1. *Strategy of Model Predictive Control*

To better understand the principle of MPC, let us have a look at the history of predictive control that started at the end of the 1970s. In the following, a list of some popular algorithms of MPC is presented:

– Model Predictive Heuristic Control (MPHC) or later known as Model Algorithmic Control (MAC) is proposed in [RIC 78]. This technique is used to control a multivariable industrial process. The control which is based on an impulse response model is computed by minimizing the tracking error.

– Dynamic Matrix Control (DMC) proposed in [CUT 80] uses the step response to predict the trajectory of the future output, and then minimizes the prediction error.

– Extended Horizon Adaptive Control (EHAC) is developed in [YDS 84] based on a parametric process model. This strategy elaborates a control sequence in order to minimize the error between the reference and the future output which is computed by solving a Diophantine equation at a period of time after the process delay.

– Extended Prediction Self-Adaptive Control (EPSAC) proposed by De Keyser and van Cauwenberghe in [DE 85] uses a discrete transfer function and a constant control signal while using a suboptimal predictor instead of solving a Diophantine equation used in EHAC method.

– Generalized Predictive Control (GPC) is one of the most popular approaches proposed in [CLA 87]. This technique is based on a Controlled Auto-Regressive Integrated Moving Average (CARIMA) model which can track both varying and constant future set points. There are some other predictive controllers based on the same idea such as: Predictive Functional Control (PFC) [RIC 78], Multi-predictor Receding Horizon Adaptive Control (MURHAC) [LEM 85], Multi-step Multivariable Adaptive Control (MUSMAR) [GRE 84] and Unified Predictive Control (UPC) [SOE 92].

– Constrained Receding Horizon Predictive Control (CRHPC) developed by Clarke and Scatollini in [CLA 91] imposes an additional terminal equality constraint on the output over a finite horizon beyond the prediction horizon.

– State-space MPC is a mature predictive control technique [MOR 94] based on the use of a state-space model. This model facilitates the stability study [KEE 88, SCO 99, MAY 00] and the generalization of MPC to multivariable systems, nonlinear systems and systems with disturbances and measurement noises. Moreover, when the system state is not available as mentioned in the previous chapters, it is more convenient to implement a state estimation (Luenberger observer, Kalman filter and set-membership estimation).

Because of the existence of uncertainties in the mathematical model, the robustness of the MPC became a key problem. The robust MPC in the presence of uncertainties and disturbances is studied by many authors [CAM 87, KOT 96, WAN 01, MAC 00, BEM 00, MAY 00, LAN 04, MAY 05, MAY 09, CHI 02, OLA 05, STO 08]. A simple way to deal with the model uncertainty is to take into account these uncertainties in the optimization problem. This means that the online optimization problem on the objective function is transformed into a *min-max* problem with constraints (the worst case of the objective function is computed on the uncertainty set and then this value of the worst case is minimized) [CAM 87]. In [KOT 96] and [WAN 01], an upper bound of the worst case is determined and then the *min-max* problem is transformed into a LMI optimization problem. To simplify the computation, [ALA 07b] proposed a suboptimal solution based on a Quadratic Programming (QP) problem. Other ways to enhance robustness are based on the deterministic MPC by ignoring the disturbances over the prediction horizon [SCO 95, CHI 02, LIM 02, MAY 05, MAY 09]. These methods require full knowledge of the state which usually cannot be reached due to measurement noises acting on real systems. In [WAN 01] and [MAY 09], the Luenberger observer is implemented to estimate the system state. In Chapters 2 and 3, the set-membership estimation has proven its efficiency in estimating the system state in the presence of bounded disturbances and measurement noises. In addition, [BEM 00] proposed output feedback predictive control based on the set-membership estimation via parallelotopes to estimate the state. Because of the efficiency of the zonotopes, in this chapter, the application of the set-membership estimation using zonotopes in the context of robust MPC is considered.

4.2. Problem formulation

Consider the general description of a Multi-Output linear discrete-time invariant system:

$$\begin{cases} \mathbf{x}_{k+1} = A\mathbf{x}_k + B\mathbf{u}_k + F\omega_k \\ \mathbf{y}_k = C\mathbf{x}_k + \mathbf{v}_k \end{cases} \qquad [4.1]$$

where $\mathbf{x}_k \in \mathbb{R}^{n_x}$ is the state of the system, $\mathbf{u}_k \in \mathbb{R}^{n_u}$ is the input vector and $\mathbf{y}_k \in \mathbb{R}^{n_y}$ is the measured output vector at time instant k. The vector $\omega_k \in \mathbb{R}^{n_x}$ represents the state perturbation vector and $\mathbf{v}_k \in \mathbb{R}^{n_y}$ represents the measurement noise vector. It is assumed that the disturbances, the measurement noises and the initial state vector are bounded by the following sets: $\omega_k \in \mathcal{W}$, $\mathbf{v}_k \in \mathcal{V}$ and $\mathbf{x}_0 \in \mathcal{X}_0$, with $\mathcal{W}$ a zonotope containing the origin, $\mathcal{V}$ a box and $\mathcal{X}_0$ a zonotope.

System [4.1] is subject to state and input constraints: $\mathbf{x}_k \in \mathcal{X}$ and $\mathbf{u}_k \in \mathcal{U}$, where $\mathcal{X}$ and $\mathcal{U}$ are two compact and convex sets containing the origin as an interior point. It is assumed that system [4.1] is stabilizable and detectable. In the following, an output feedback control based on invariant tube of uncertain trajectories [MAY 09] is presented. This control law allows us to improve the performance of the control system, and it can also guarantee the stability of the system and the feasibility of the control law.

4.3. Tube-based output feedback Model Predictive Control design

This section presents an MPC technique based on feedback control. In feedback MPC, the decision variable is a control policy, which is a sequence of control laws. The idea of the Tube-based Model Predictive Control (TMPC) [MAY 00] is to construct a tube of all the possible trajectories around the nominal trajectory ensuring robust constraints satisfaction and stability guarantees of the controlled system, using invariant set techniques. In addition, the robustness of the predictive control laws under disturbances and uncertainties has been addressed by different authors [LEE 97, BEM 98, BEM 99]. During the past few years, the TMPC has been investigated by many authors [LAN 04, CHI 02, MAY 05, LIM 02, LIM 07, MAY 09]. In this section, the TMPC developed in [MAY 05] and [MAY 09] will be summarized. In this approach, the optimization problem of the true system state is replaced by the nominal state, that is the state of the nominal system which is not influenced by disturbances and measurement noises. In addition, the nominal state is considered as a decision variable in the optimal control problem solved online. This consideration allows us to facilitate the proof of stability and attractivity of the terminal set. Moreover, the complexity of the controller is similar to that required for nominal MPC. This technique is developed for linear time invariant systems with the hypothesis that the system states are known [MAY 05]. When the system state vector is unknown, the state is estimated using an observer. Then the estimation error is considered unknown but bounded uncertainty which can be taken into account in the control law. In [MAY 09], the Luenberger observer is used to estimate the system state. The gain matrix of the Luenberger observer plays an important role for the performances of the observer, and thus the global performance of the controlled system. The main difficulty comes from the fact that the choice of this gain matrix is not unique. For this reason, in this section, the P-radius-based set-membership estimation presented in the previous chapter is chosen to replace the Luenberger observer while combined with the TMPC. Using this zonotopic set-membership estimation, the bound of the estimation error is minimized and, thus, the performance of the controlled system is improved.

The nominal system is now defined as:

$$\bar{\mathbf{x}}_{k+1} = A\bar{\mathbf{x}}_k + B\bar{\mathbf{u}}_k \tag{4.2}$$

Note that for pedagogical reasons, the same notations as in [MAY 09] have been used. Now consider system [4.1]; due to the presence of the disturbances and the measurement noises, the system state is estimated using the zonotopic guaranteed state estimation presented in the previous chapter. The center of the zonotopic guaranteed state estimation at time instant k is considered as the estimation state at time instant k: $\hat{\mathbf{x}}_k = \hat{\mathbf{p}}_k$. Similar to the method presented in the previous section with the added control law $\mathbf{u}_k$, the dynamics of the estimation state is determined by the following equation:

$$\begin{aligned} \hat{\mathbf{x}}_{k+1} &= A\hat{\mathbf{x}}_k + B\mathbf{u}_k + \Lambda(\mathbf{y}_{k+1} - \hat{\mathbf{y}}_{k+1}) \\ &= A\hat{\mathbf{x}}_k + B\mathbf{u}_k + \Lambda(\mathbf{y}_{k+1} - C(A\hat{\mathbf{x}}_k + B\mathbf{u}_k)) \end{aligned} \tag{4.3}$$

Replacing $\mathbf{y}_{k+1} = C\mathbf{x}_{k+1} + \mathbf{v}_{k+1}$ with $\mathbf{x}_{k+1} = A\mathbf{x}_k + B\mathbf{u}_k + \omega_k$, this is equivalent to:

$$\begin{aligned} \hat{\mathbf{x}}_{k+1} &= A\hat{\mathbf{x}}_k + B\mathbf{u}_k + \Lambda(CA\mathbf{x}_k + CB\mathbf{u}_k + C\omega_k + \mathbf{v}_{k+1} \\ &\qquad - CA\hat{\mathbf{x}}_k - CB\mathbf{u}_k) \\ &= A\hat{\mathbf{x}}_k + B\mathbf{u}_k + \Lambda CA(\mathbf{x}_k - \hat{\mathbf{x}}_k) + \Lambda C\omega_k + \Lambda\mathbf{v}_{k+1} \end{aligned} \tag{4.4}$$

Denote the estimation error as $\tilde{\mathbf{x}}_k = \mathbf{x}_k - \hat{\mathbf{x}}_k$; from equations [4.1] and [4.4], the dynamics of the estimation error is described by:

$$\tilde{\mathbf{x}}_{k+1} = (I - \Lambda C)A\tilde{\mathbf{x}}_k + \omega_k^e \tag{4.5}$$

with $\omega_k^e = (I - \Lambda C)\omega_k - \Lambda\mathbf{v}_{k+1}$. From the hypotheses $\omega_k \in \mathcal{W}$ and $\mathbf{v}_k \in \mathcal{V}$, the following expression holds:

$$\omega_k^e \in \mathcal{W}^e = (I - \Lambda C)\mathcal{W} \oplus (-\Lambda\mathcal{V}) \tag{4.6}$$

Consider that the estimation error at time instant k is bounded by a set $\mathcal{S}_k^e$, equation [4.5] leads to the following recursive expression:

$$\tilde{\mathbf{x}}_{k+1} \in \mathcal{S}_{k+1}^e = (I - \Lambda C)A\mathcal{S}_k^e \oplus \mathcal{W}^e \qquad [4.7]$$

In order to ensure the stability of the controlled system, the estimation error must be taken into account in the control law. The main idea behind the design of the TMPC control law comes from the non-increasing property of the set $\mathcal{S}_{k+1}^e$ which bounds the future estimation error. In the following, this property is provided by the robust positive invariance [BLA 99] of the set $\mathcal{S}_{k+1}^e$, which is defined in the following.

DEFINITION 4.1.– [BLA 99] – A set Ω is called robust positive invariant set for the system $\mathbf{x}_{k+1} = A\mathbf{x}_k + \omega_k$ subject to the constraints $(\mathbf{x}_k, \omega_k) \in (\mathcal{X}, \mathcal{W})$, if $A\mathbf{x}_k + \omega_k \in \Omega$ and $\Omega \in \mathcal{X}$, $\forall \omega_k \subseteq \mathcal{W}$ and $\forall \mathbf{x}_k \in \Omega$. Thus, the maximal robust positive invariant set Ω_{max} for the same system is the smallest robust positive invariant set that contains all the robust positive invariant sets of this system. The minimal robust positive invariant set Ω_{min} of the same system is the robust positive invariant set which is contained in any robust positive invariant set of this system.

PROPOSITION 4.1.– [MAY 09] – If the set $\mathcal{S}_0^e$ is robust positively invariant for system [4.5], then the sets $\mathcal{S}_k^e$ are robust positively invariant for the system [4.5], too. The set sequence $\{\mathcal{S}_k^e\}$ is monotonic non-increasing and converges in the Hausdorff metric to a robust positively invariant set $\mathcal{S}_\infty^e = (I - \Lambda C)A\mathcal{S}_\infty^e \oplus \mathcal{W}^e$.

Note that the original idea behind the use of the zonotopic set-membership estimation in this book in comparison with the use of Luenberger observer in [MAY 09] is that the matrix Λ is computed to optimize the size of the set $\mathcal{S}_\infty^e$.

The observer equation can be rewritten as:

$$\hat{\mathbf{x}}_{k+1} = A\hat{\mathbf{x}}_k + B\mathbf{u}_k + \omega_k^{co} \quad [4.8]$$

with $\omega_k^{co} = \Lambda(CA\tilde{\mathbf{x}}_k + C\omega_k + \mathbf{v}_{k+1})$. As $\tilde{\mathbf{x}}_k \in \mathcal{S}_k^e$, the following expression is true:

$$\omega_k^{co} \in \mathcal{W}_k^{co} = \Lambda CA\mathcal{S}_k^e \oplus \Lambda C\mathcal{W} \oplus \Lambda\mathcal{V} \quad [4.9]$$

Because the set sequence $\{\mathcal{S}_k^e\}$ is monotonic non-increasing as presented in Proposition 4.1 and $\mathcal{W}_k^{co}$ is linearly dependent on $\mathcal{S}_k^e$, thus, the set sequence $\{\mathcal{W}_k^{co}\}$ is monotonic non-increasing, too.

Now consider the nominal system which is not affected by disturbances:

$$\underline{\mathbf{x}}_{k+1} = A\underline{\mathbf{x}}_k + B\underline{\mathbf{u}}_k \quad [4.10]$$

To counteract the disturbances, the trajectory of the real system is desired to lie close to the nominal system trajectory. If the nominal system is steered to the origin, then the center of the zonotopic state estimation as well as the real state of the system are bounded by a compact set. In order to have all the possible trajectories inside a tube around the nominal trajectory, a control law $\mathbf{u}_k$ can be defined as:

$$\mathbf{u}_k = \underline{\mathbf{u}}_k + K(\hat{\mathbf{x}}_k - \underline{\mathbf{x}}_k) \quad [4.11]$$

where $\underline{\mathbf{u}}_k$ is the control law applied to the nominal system [4.10] at time instant k.

Combining [4.8] and [4.10], the error between the estimation state and the nominal state denoted by $\mathbf{e}_k = \hat{\mathbf{x}}_k - \underline{\mathbf{x}}_k$ satisfies the difference equation:

$$\begin{aligned}\mathbf{e}_{k+1} &= \hat{\mathbf{x}}_{k+1} - \underline{\mathbf{x}}_{k+1} \\ &= (A\hat{\mathbf{x}}_k + B\mathbf{u}_k + \omega_k^{co}) - (A\underline{\mathbf{x}}_k + B\underline{\mathbf{u}}_k)\end{aligned} \quad [4.12]$$

Replacing $\mathbf{u}_k$ by [4.11], $\mathbf{e}_{k+1}$ can be further written as:

$$\mathbf{e}_{k+1} = (A + BK)\mathbf{e}_k + \boldsymbol{\omega}_k^{co} \qquad [4.13]$$

The gain matrix K is chosen such that the matrix $A + BK$ is Schur stable. Therefore, if at time instant k, $\mathbf{e}_k$ lies in the set $\mathcal{S}_k^{co}$, then $\mathbf{e}_{k+1}$ lies in the set:

$$\mathcal{S}_{k+1}^{co} = (A + BK)\mathcal{S}_k^{co} \oplus \mathcal{W}_k^{co} \qquad [4.14]$$

PROPOSITION 4.2.– [MAY 09] – If $\mathbf{e}_0$ belongs to an initial set $\mathcal{S}_0^{co}$, then the set sequence $\{\mathcal{S}_k^{co}\}$ converges to a robust positive invariant set $\mathcal{S}_\infty^{co}$ which satisfies $\mathcal{S}_\infty^{co} = (A + BK)\mathcal{S}_\infty^{co} \oplus \mathcal{W}_\infty^{co}$. If $\mathcal{S}_0^e$ is robust positive invariant for system [4.5] and $\mathcal{S}_0^{co}$ is robust positive invariant for system [4.13] with $\omega_0^{co} \in \mathcal{W}_0^{co}$, then the set sequence $\{\mathcal{S}_k^{co}\}$ is robust positive invariant for system [4.13] with $\omega_k^{co} \in \mathcal{W}_k^{co}$ and monotonic non-increasing ($\mathcal{S}_{k+1}^{co} \subseteq \mathcal{S}_k^{co}$).

Based on the procedure proposed in [MAY 09], let us define $\mathcal{S}_k = \mathcal{S}_k^e \oplus \mathcal{S}_k^{co}$. Because $\mathcal{S}_k^e$ converges to $\mathcal{S}_\infty^e$ and $\mathcal{S}_\infty^{co}$ converges to $\mathcal{S}_\infty^{co}$, then $\mathcal{S}_k$ tends to $\mathcal{S}_\infty = \mathcal{S}_\infty^e \oplus \mathcal{S}_\infty^{co}$.

In this context, the robust TMPC can be summarized as follows. At time instant k a state estimation set is computed and a nominal optimal control problem is solved online. The cost function for the nominal system is defined as:

$$V_N(\underline{\mathbf{x}}_k, \underline{\mathbf{u}}) = \frac{1}{2}V_f(\underline{\mathbf{x}}_{k+N}) + \sum_{i=0}^{N-1} \frac{1}{2} l(\underline{\mathbf{x}}_{k+i}, \underline{\mathbf{u}}_{k+i}) \qquad [4.15]$$

where N is the prediction horizon and $\bar{\mathbf{u}}$ is the control sequence:

$$\underline{\mathbf{u}} = \{\underline{\mathbf{u}}_k, \underline{\mathbf{u}}_{k+1}, \ldots, \underline{\mathbf{u}}_{k+N-1}\} \qquad [4.16]$$

The stage cost function $l(\underline{\mathbf{x}}_k, \underline{\mathbf{u}}_k)$ and the terminal cost function $V_f(\underline{\mathbf{x}}_{k+N})$ are defined by:

$$\begin{cases} l(\underline{\mathbf{x}}_k, \underline{\mathbf{u}}_k) = \frac{1}{2}(\underline{\mathbf{x}}_k^T Q \underline{\mathbf{x}}_k + \underline{\mathbf{u}}_k^T R \underline{\mathbf{u}}_k) \\ V_f(\underline{\mathbf{x}}_{k+N}) = \frac{1}{2}\mathbf{x}_{k+N}^T P_f \underline{\mathbf{x}}_{k+N} \end{cases} \qquad [4.17]$$

where P_f, Q, R are positive definite matrices. With these notations, the time varying constraints at current time instant k are:

$$\begin{cases} \underline{\mathbf{u}}_{k+i} \in \underline{\mathcal{U}}_{k+i},\ i = 0, \ldots, N-1 \\ \underline{\mathbf{x}}_{k+i} \in \underline{\mathcal{X}}_{k+i},\ i = 0, \ldots, N-1 \\ \underline{\mathbf{x}}_{k+N} \in \underline{\mathcal{X}}_f \end{cases} \qquad [4.18]$$

with $\underline{\mathcal{U}}_{k+i} = \mathcal{U} \ominus K\mathcal{S}_{k+i}^{co}$ (this tight constraint comes from equation [4.11]), $\underline{\mathcal{X}}_{k+i} = \mathcal{X} \ominus \mathcal{S}_{k+i}$ (this constraint comes from the fact that $\mathbf{x}_k = \underline{\mathbf{x}}_k + \tilde{\mathbf{x}}_k + \mathbf{e}_k$, with conditions [4.7] and [4.14]).

To ensure the feasibility and the stability of this control law, the following assumptions are assumed [MAY 09].

ASSUMPTION 4.1.– Consider $\mathcal{S}_0 = \mathcal{S}_0^{co} \oplus \mathcal{S}_0^{e} \subset \mathcal{X}$ and $K\mathcal{S}_0^{co} \subset \mathcal{U}$.

This condition is assumed in order to ensure that the initial condition satisfies the constraints.

The terminal cost V_f and the terminal constraint set $\underline{\mathcal{X}}_f$ are assumed to satisfy the stabilizing condition [MAY 00] (assumptions 4.2 and 4.3).

ASSUMPTION 4.2.– $\underline{\mathcal{X}}_f$ is a proper C-set, is positive invariant for $\underline{\mathbf{x}}_{k+1} = (A + BK)\underline{\mathbf{x}}_k$ and satisfies $\underline{\mathcal{X}}_f \subseteq \underline{\mathcal{X}}_N$ and $K\underline{\mathcal{X}}_f \subseteq \underline{\mathcal{U}}_N$.

ASSUMPTION 4.3.– $V_f(.)$ is a local control Lyapunov function for $\underline{\mathbf{x}}_{k+1} = (A + BK)\underline{\mathbf{x}}_k$ for all $\underline{\mathbf{x}} \in \underline{\mathcal{X}}_f$. There exist constants

c_1, $c_2 > 0$ such that $c_1|\underline{\mathbf{x}}_k|^2 \leq V_f(\underline{\mathbf{x}}_k) \leq c_2|\underline{\mathbf{x}}_k|^2$ and $V_f((A + BK)\underline{\mathbf{x}}_k) + l(\underline{\mathbf{x}}_k, K\underline{\mathbf{x}}_k) \leq V_f(\underline{\mathbf{x}}_k)$. This means that the Lyapunov function is decreased at the next sampling time.

Denote the set of admissible control sequences at time instant k, with the nominal state $\underline{\mathbf{x}}_k$:

$$U_N(\underline{\mathbf{x}}_k) = \{\underline{\mathbf{u}} : \underline{\mathbf{u}}_{k+i} \in \underline{\mathcal{U}}_{k+i}, \underline{\mathbf{x}}_{k+i} \in \underline{\mathcal{X}}_{k+i}, \underline{\mathbf{x}}_{k+N} \in \underline{\mathcal{X}}_f, \\ i = 0, \dots, N-1\} \qquad [4.19]$$

Then the nominal optimal control problem is:

$$V_N^0(\underline{\mathbf{x}}_k) = \min_{\underline{\mathbf{u}}}\{V_N(\underline{\mathbf{x}}_k, \underline{\mathbf{u}}) : \underline{\mathbf{u}} \in U_N(\underline{\mathbf{x}}_k)\} \qquad [4.20]$$

At each time instant k, the feasible domain of $\underline{x}_k$ is given by:

$$\underline{\mathcal{X}}_N(k) = \{\underline{\mathbf{x}}_k : U_N(\underline{\mathbf{x}}_k) \neq 0\} \qquad [4.21]$$

which leads to $\underline{\mathcal{X}}_N(k) \subseteq \underline{\mathcal{X}}_N(k+1)$.

PROPOSITION 4.3.– [MAY 09] – There exist two constants c_1 and c_2 such that $\forall \underline{\mathbf{x}}_k \in \underline{\mathcal{X}}_N(k)$, $k \in \mathbb{N}^+$ the following expressions hold:

1) $c_1|\underline{\mathbf{x}}_k| \leq V_N^0(\underline{\mathbf{x}}_k) \leq c_2|\underline{\mathbf{x}}_k|$;

2) $V_N^0(\underline{\mathbf{x}}_{k+1}) \leq V_N^0(\underline{\mathbf{x}}_k) - c_1|\underline{\mathbf{x}}|^2$.

This proposition establishes the exponential stability at the origin for the nominal system.

Now consider the optimization control problem solved online:

$$V_N^*(\hat{\mathbf{x}}_k) = \min_{\underline{\mathbf{x}}_k, \underline{\mathbf{u}}}\{V_N(\underline{\mathbf{x}}_k, \underline{\mathbf{u}}) : \underline{\mathbf{u}} \in \mathcal{U}_N(\underline{\mathbf{x}}_k), \hat{\mathbf{x}}_k \in \underline{\mathbf{x}}_k \oplus \mathcal{S}_k^{co}\} \qquad [4.22]$$

Let us consider the solution of this optimization problem:

$$(\underline{\mathbf{x}}_k^*(\hat{\mathbf{x}}_k), \underline{\mathbf{u}}^*(\hat{\mathbf{x}}_k)) = \arg\min_{\underline{\mathbf{x}}_k, \underline{\mathbf{u}}}\{V_N(\underline{\mathbf{x}}_k, \underline{\mathbf{u}}) : \underline{\mathbf{u}} \in U_N(\underline{\mathbf{x}}_k), \\ \hat{\mathbf{x}}_k \in \underline{\mathbf{x}}_k \oplus \mathcal{S}_k^{co}\} \qquad [4.23]$$

then, using [4.11], the control law applied to the system is obtained:

$$\kappa_N(\hat{\mathbf{x}}_k) = \hat{\mathbf{u}}_k^*(\hat{\mathbf{x}}_k) + K(\hat{\mathbf{x}}_k - \underline{\mathbf{x}}^*(\hat{\mathbf{x}}_k)) \qquad [4.24]$$

with $\hat{\mathbf{u}}_k^*(\hat{\mathbf{x}}_k)$ the first element of the sequence $\underline{\mathbf{u}}^*(\hat{\mathbf{x}}_k)$.

Using this control law, it can be proved that $(\mathbf{x}_k, \hat{\mathbf{x}}_k)$ is robustly steered to $\mathcal{S}_\infty \times \mathcal{S}_\infty^{co}$, exponentially fast satisfying all constraints [MAY 09].

This robust TMPC control law is illustrated by two examples.

EXAMPLE 4.1.– Consider a second-order system:

$$\begin{cases} \mathbf{x}_{k+1} = \begin{bmatrix} 1 & 1.1 \\ 0 & 1 \end{bmatrix} \mathbf{x}_k + \begin{bmatrix} 1 \\ 1 \end{bmatrix} u_k + \boldsymbol{\omega}_k \\ y_k = \begin{bmatrix} -2 & 1 \end{bmatrix} \mathbf{x}_k + v_k \end{cases} \qquad [4.25]$$

The disturbances and measurement noises are assumed to be bounded $(\omega, v) \in \mathcal{W} \times \mathcal{V}$, where $\mathcal{W} = \{\omega \in \mathbb{R}^2 \colon \|\omega\|_\infty \leq 0.1\}$ and $\mathcal{V} = \{v \in \mathbb{R} \colon |v| \leq 0.05\}$. The state and control constraints are $(\mathbf{x}_k, u_k) \in \mathcal{X} \times \mathcal{U}$, where $\mathcal{X} = \{\mathbf{x} \in \mathbb{R}^2 : x_1 \in [-50, 3], x_2 \in [-50, 3]\}$ and $\mathcal{U} = \{u \in \mathbb{R} \colon |u| \leq 9\}$. This system must be stabilized around the origin while respecting the aforementioned state and control constraints.

The weighting matrices in the cost function are $Q = I_2$ and $R = 0.01$. The terminal cost $V_f(\underline{\mathbf{x}})$ is the value function $\underline{\mathbf{x}}^T P_f \underline{\mathbf{x}}$ for the unconstrained optimal control problem for the nominal system $\underline{\mathbf{x}}_{k+1} = A\underline{\mathbf{x}}_k + B\underline{u}_k$. The control signal $\underline{u}_k = K\underline{\mathbf{x}}_k$ is the associated Linear Quadratic Regulator (LQR) control $K = \begin{bmatrix} -0.6029 & -1.0567 \end{bmatrix}$. The initial sets $\mathcal{S}_0^e$, $\mathcal{S}_0^{co}$ are computed using a recursive algorithm to compute an outer approximation of the minimal invariant set [RAK 05]. The initial state belongs to the zonotope $\mathcal{S}_0^e$) centered in $\begin{bmatrix} 0 \\ -15 \end{bmatrix}$. The prediction horizon is chosen $N = 13$ such that the optimization is feasible at the initial time instant.

The terminal constraint set $\underline{\mathcal{X}}_f$ (the black set depicted in Figure 4.4) is the maximal positive invariant set for the system $\underline{\mathbf{x}}_{k+1} = (A + BK)\underline{\mathbf{x}}_k$ under the tighter constraints $\underline{\mathcal{X}}_N = \mathcal{X} \ominus \mathcal{S}_N$ and $\underline{\mathcal{U}}_N = \mathcal{U} \ominus K\mathcal{S}_N^{co}$ obtained using the recursive algorithm proposed by [GIL 91]. This algorithm is based on a search of the maximal value of t such that $(A + BK)^{t+1}\underline{\mathbf{x}} \in \underline{\mathcal{X}}_N$ is subject to the constraint $(A + BK)^k\underline{\mathbf{x}} \in \underline{\mathcal{X}}_N$, with $k = 0, \ldots, t$.

Figure 4.2 and its zoom (Figure 4.3) compare the state estimation sets of three approaches: the segment minimization, the volume minimization and the P-radius minimization. The zonotopic guaranteed state estimation obtained by P-radius minimization is non-increasing in time. These figures confirm the trade-off of the P-radius minimization between the segment minimization and the volume minimization.

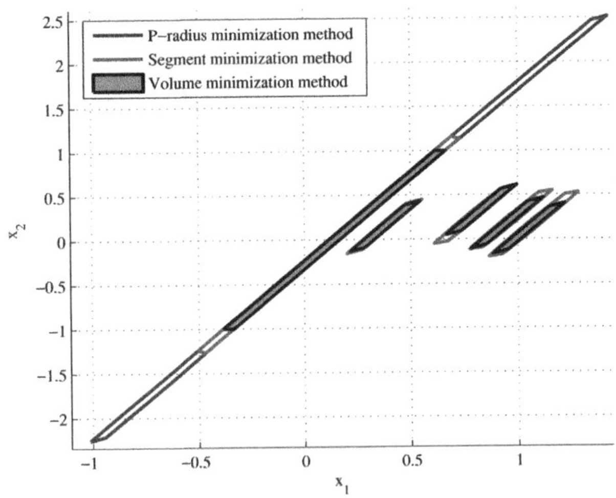

Figure 4.2. *Evolution of the zonotopic state estimation set. For a color version of this figure, see www.iste.co.uk / stoica / zonotopes.zip*

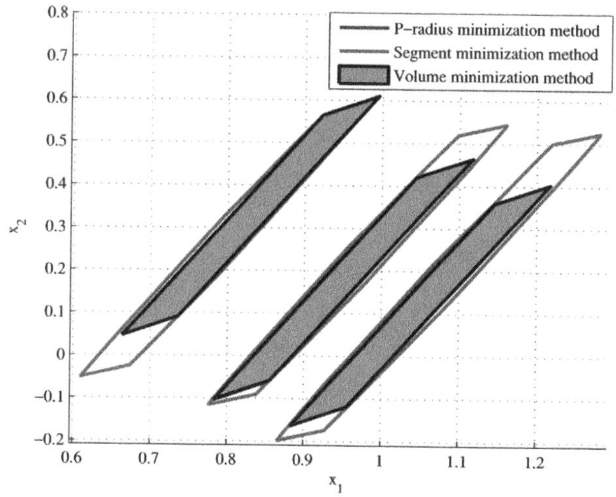

Figure 4.3. *Zoom of the evolution of the zonotopic state estimation set. For a color version of this figure, see www.iste.co.uk / stoica / zonotopes.zip*

Figure 4.4 shows the tube trajectory of the system. The largest zonotope is the set $\bar{\mathbf{x}}^*(\hat{\mathbf{x}}_k, k) \oplus \mathcal{S}_k$. The real state is guaranteed to belong to this tube section centered at the nominal state. The smaller zonotope is the set $\bar{\mathbf{x}}^*(\hat{\mathbf{x}}_k, k) \oplus \mathcal{S}_k^{co}$, the state estimation is guaranteed to belong to this set. The smallest zonotope is the guaranteed state estimation set $\hat{\mathbf{x}}_k \oplus \mathcal{S}_k^e$. Denote that the largest zonotope which is the section of the tube trajectory at each time instant is the Minkowski sum of the smallest zonotope and the smaller zonotope $\mathcal{S}_k = \mathcal{S}_k^{co} \oplus \mathcal{S}_k^e$. Because of the non-increase in time of the guaranteed state estimation, the section of tube is non-increasing in time. This trajectory is put in the box of the state constraint (Figure 4.5) in order to illustrate that the system respects the constraint. Figure 4.6 shows the stability of this output feedback system respecting the constraints. Because of the presence of the uncertainty, the system state does not converge to the origin but converges to a set containing the origin.

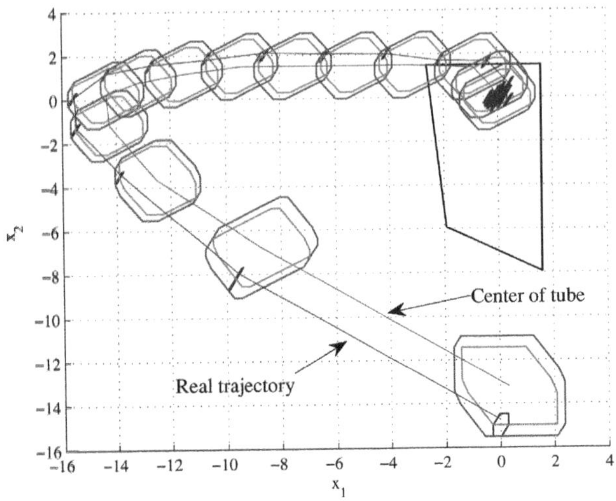

Figure 4.4. *Tube trajectory of the closed-loop system. For a color version of this figure, see www.iste.co.uk / stoica / zonotopes.zip*

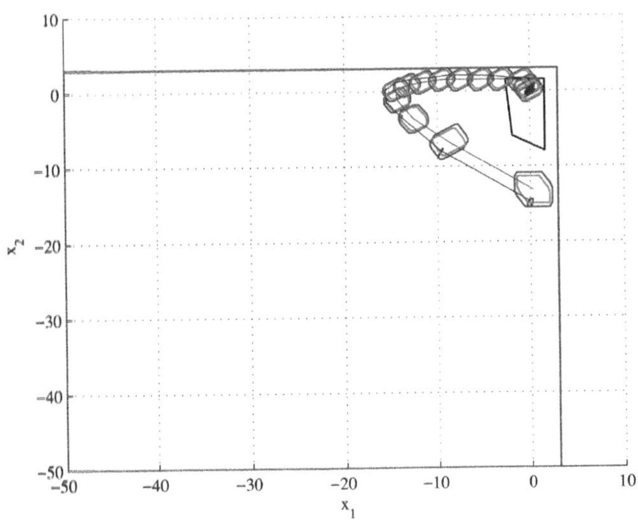

Figure 4.5. *Tube trajectory of the closed-loop system fulfilling the state constraints. For a color version of this figure, see www.iste.co.uk / stoica / zonotopes.zip*

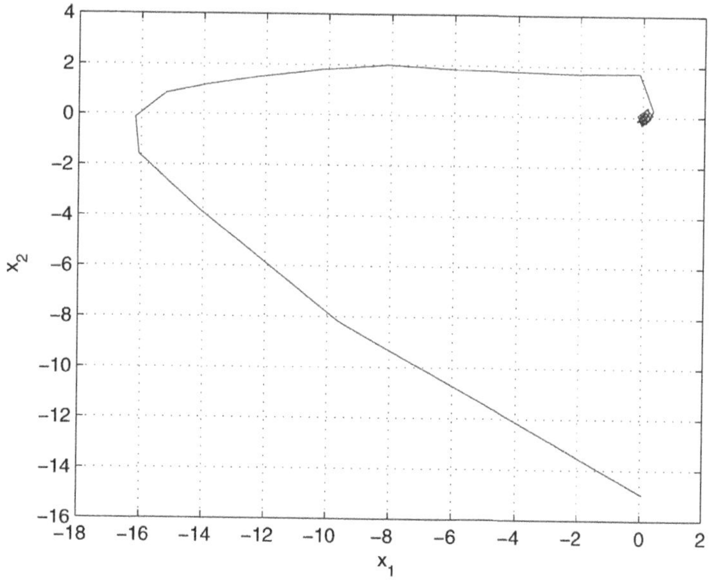

Figure 4.6. *Closed-loop response of the system*

4.4. Application on the magnetic levitation system

In this section, the control and estimation approaches proposed in this chapter are applied to the model of a real system: the magnetic levitation system. The goal is to test the zonotopic set-membership estimation and the association between this estimation and TMPC. First, a general description and the mathematical model of the magnetic levitation system are proposed. Second, based on this model, the MPC law is built using the zonotopic set-membership estimation. The simulation results show that the predictive control method based on the zonotopic set-membership estimation can stabilize this system fulfilling the considered constraints, despite the disturbances and measurement noises acting on the system.

4.4.1. *System description*

This system is composed of a mobile iron pendulum in a vertical magnetic field created by a fixed electromagnet (Figure 4.7). This electromagnet is supplied by a variable current which allows us to vary the magnetic force, and thus vary the vertical position of the pendulum. This system is assumed to have a perfect radial symmetry.

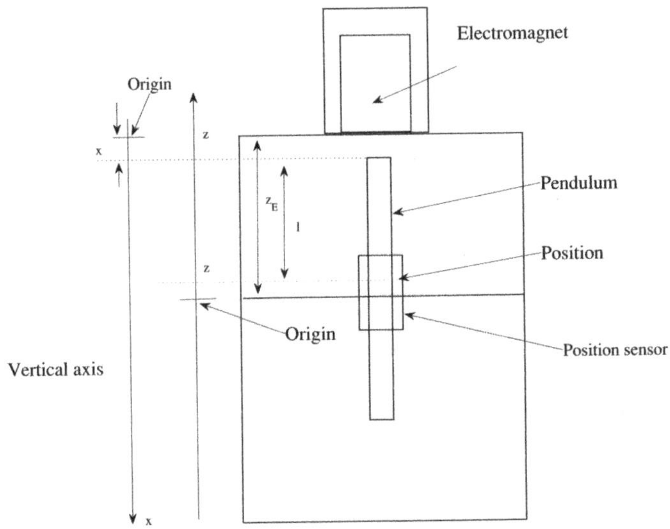

Figure 4.7. *Magnetic levitation system*

The control goal is to stabilize the vertical position of the pendulum around the equilibrium point. Figure 4.8 illustrates a general schematic block diagram of the closed-loop system with z the position of the pendulum relative to the sensor position center in an absolute reference frame and i the current given by the actuator. In the following section, the model of the magnetic levitation without the actuator is considered in order to simplify the computation and to maintain a 2D simple visualization.

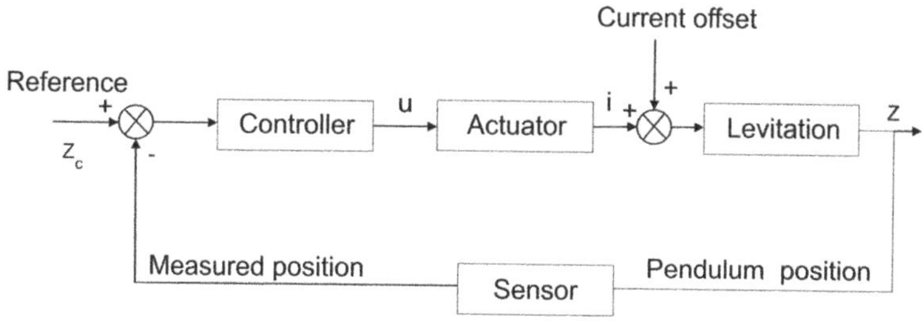

Figure 4.8. *Block diagram of the closed-loop system*

The force applied to the pendulum created by the electromagnet has the following form:

$$F_m(t) = \frac{ci(t)}{x^2(t)} \quad [4.26]$$

where i is the current in the electromagnet, x is the distance between the pendulum and the electromagnet and c is a constant. Using the fundamental relation of the pendulum dynamics, the movement of the pendulum can be described by the following differential equation:

$$-m\ddot{x}(t) = -mg + \frac{ci(t)}{x^2(t)} \quad [4.27]$$

where m is the pendulum mass and g is the gravitational constant. Replacing $x(t) = x_0 - z(t)$ in equation [4.27], where $x_0 = z_E - l$ (Figure 4.7) leads to:

$$m\ddot{z}(t) = -mg + \frac{ci(t)}{x_0^2(1 - \frac{z(t)}{x_0})^2} \quad [4.28]$$

Consider a small variation of $z(t)$ around the origin (with $|z(t)| << x_0$) and $i(t)$ around i_0 (with i_0 the current at the equilibrium point, $i(t) = i_0 + i_1(t)$ and $|i_1(t)| << i_0$ a small variation around i_0). Keeping only the first-order terms, equation [4.28] is then approximated by:

$$\begin{aligned} m\ddot{z}(t) &\approx -mg + \tfrac{ci(t)}{x_0^2}(1 + \tfrac{2z(t)}{x_0}) \\ &\approx -mg + \tfrac{ci(t)}{x_0^2} + \tfrac{2ci(t)z(t)}{x_0^3} \end{aligned} \quad [4.29]$$

Writing equation [4.28] at the equilibrium point ($z(t) = 0$) leads to:

$$\frac{ci_0}{x_0^2} - mg = 0 \quad [4.30]$$

Replacing [4.30] and $i(t) = i_0 + i_1(t)$ in equation [4.29], the following expression is obtained:

$$\frac{ci_0}{gx_0^2}\ddot{z}(t) = \frac{ci_1(t)}{x_0^2} + \frac{2ci(t)z(t)}{x_0^3} \quad [4.31]$$

Because $i_1(t) << i_0$, the following approximation can be done $i(t) \approx i_0$. Replacing this approximation in equation [4.31] leads to:

$$\frac{ci_0}{gx_0^2}\ddot{z}(t) \approx \frac{ci_1(t)}{x_0^2} + \frac{2ci_0z(t)}{x_0^3} \quad [4.32]$$

or multiplying [4.32] by $\frac{x_0^2}{c}$ leads to:

$$\frac{i_0}{g}\ddot{z}(t) = i_1(t) + \frac{2i_0}{x_0}z(t) \quad [4.33]$$

Based on this differential equation [4.33], the dynamics of the magnetic levitation in continuous time can be described as the following state-space representation:

$$\begin{cases} \dot{\mathbf{x}} = \begin{bmatrix} 0 & 1 \\ \frac{2g}{x_0} & 0 \end{bmatrix} \mathbf{x} + \begin{bmatrix} 0 \\ \frac{g}{i_0} \end{bmatrix} u \\ y = \begin{bmatrix} 1 & 0 \end{bmatrix} \mathbf{x} \end{cases} \qquad [4.34]$$

where $\mathbf{x} = \begin{bmatrix} x_1 & x_2 \end{bmatrix}$ correspond to the position z and the speed of the pendulum $\dot{z}$, respectively, u corresponds to the electromagnet current $i_1(t)$, y is the position measurement, $g = 9.81\,\mathrm{m/s^2}$, $x_0 = 0.019\,\mathrm{m}$ and $i_0 = 0.436\,\mathrm{A}$. As the magnetic levitation system is a nonlinear system, this model is only a local model linearized around the origin. The eigenvalues of the open-loop evolution matrix are 32.1346 and -32.1346. This means that the open-loop system is unstable, which confirms the unstable physical behavior of the magnetic levitation system. Note that even if the position is measured by a position sensor, due to the disturbances and measurement noises, the zonotopic set-membership estimation is implemented to estimate not only the position but also the speed of the pendulum.

4.4.2. *Control problem*

As system [4.34] is unstable in open loop, the control problem consists of stabilizing this unstable system around the origin under the constraints $(\mathbf{x}, u) \in \mathcal{X} \times \mathcal{U}$, with $\mathcal{X} = \{\mathbf{x} \in \mathbb{R}^2 : x_1 \in [-0.5; 0.5]\mathrm{m}, x_2 \in [-10; 10]\mathrm{m/s}\}$ and $\mathcal{U} = \{u \in \mathbb{R} : |u| \leq 5\mathrm{A}\}$. Equation [4.34] is discretized using the zero-order hold on the inputs with the sample time $T_s = 0.1\,\mathrm{s}$. The obtained linearized discrete-time invariant system is the following:

$$\begin{cases} \mathbf{x}_{k+1} = \begin{bmatrix} 12.4526 & 0.3863 \\ 398.8660 & 12.4526 \end{bmatrix} \mathbf{x}_k + \begin{bmatrix} 0.2495 \\ 8.6909 \end{bmatrix} u_k + \begin{bmatrix} 0.001 \\ 0.01 \end{bmatrix} \omega_k \\ y_k = \begin{bmatrix} 1 & 0 \end{bmatrix} \mathbf{x}_k + v_k \end{cases} \qquad [4.35]$$

where ω_k and v_k are added in order to model the disturbances and the measurement noises: $\omega \in \mathcal{W} = \{\omega \in \mathbb{R} : |\omega| \leq 1\}$, $v \in \mathcal{V} = \{v \in \mathbb{R} : |v| \leq 0.05\,\mathrm{m}\}$.

This control problem is solved using the TMPC based on the P-radius zonotopic set-membership estimation procedure.

The first simulation result consists of comparison of the P-radius-based method of guaranteed state estimation obtained by solving the optimization problem [3.22] with the volume minimization method and the segment minimization method [ALA 05] at time instants $k = 1$ and $k = 2$ (Figure 4.9 and its zoom Figure 4.10). In this example, these three methods provide similar estimation results. The minimization of the P-radius method has the same computation time as the segment minimization method and is smaller than the volume minimization method as shown in Table 4.1.

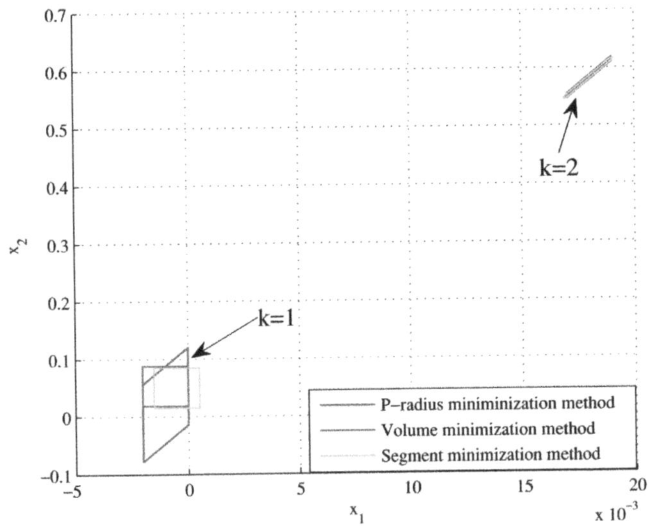

Figure 4.9. *Comparison of the zonotopic guaranteed state estimation obtained by different approaches. For a color version of this figure, see www.iste.co.uk / stoica / zonotopes.zip*

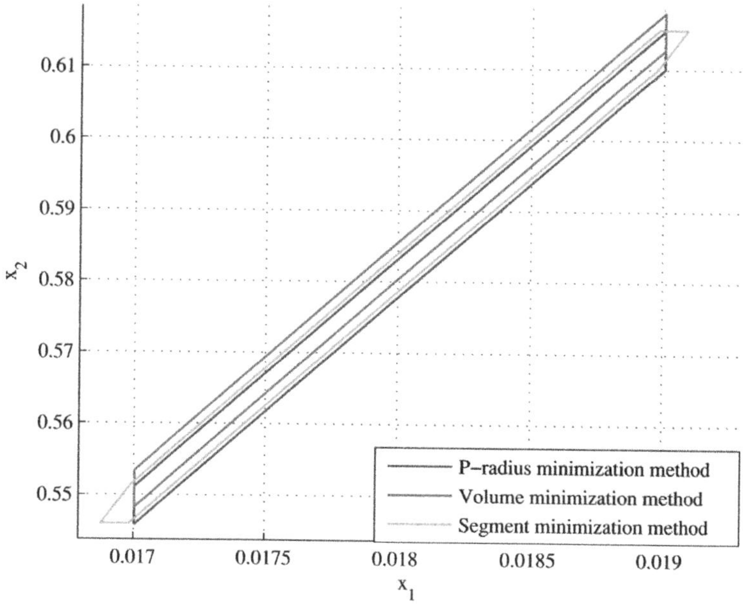

Figure 4.10. *Zoom of the comparison of the zonotopic guaranteed state estimation obtained by different approaches. For a color version of this figure, see www.iste.co.uk/stoica/zonotopes.zip*

Algorithm	Time (seconds)
Segment minimization	0.0468
P-radius minimization (without off-line optimization included)	0.0312
P-radius minimization (with off-line optimization [3.22] included)	0.7488
Volume minimization	7.8469

Table 4.1. *Total computation time of the estimation problem of the magnetic levitation system after 50 time instants*

These estimation results will be further used within the Tube Model Predictive Control law. The TMPC based on the zonotopic set-membership estimation is used to stabilize this system as follows. The weighting matrices in the cost function [4.15] are chosen as identity matrices $Q = R = I_2$. The terminal cost $V_f(\underline{\mathbf{x}})$ is the value function $\underline{\mathbf{x}}^T P_f \underline{\mathbf{x}}$ for the unconstrained optimal control problem for the nominal system $\underline{\mathbf{x}}_{k+1} = A\underline{\mathbf{x}}_k + B\underline{u}_k$ where $\underline{u}_k = K\underline{\mathbf{x}}_k$ is the associated LQR. The initial sets $\mathcal{S}_0^e$ (the guaranteed state estimation at time instant $k = 0$) and $\mathcal{S}_0^{co}$ (the set contains the error between the estimation state and the nominal state at time instant $k = 0$) are computed using the result in [RAK 05] as presented in the previous section. The initial state belongs to the zonotope $\mathcal{S}_0^e$ centered in $\begin{bmatrix} 0.1 \\ -6.5 \end{bmatrix}$. The prediction horizon is chosen $N = 3$ such that the optimization is feasible at the initial time instant. The terminal constraint set $\underline{\mathcal{X}}_f$ (the black set depicted in Figure 4.11) is the maximal positive invariant set for the system $\underline{\mathbf{x}}_{k+1} = (A + BK)\underline{\mathbf{x}}_k$ under the tighter constraints $\underline{\mathcal{X}}_N = \mathcal{X} \ominus \mathcal{S}_N$ and $\underline{\mathcal{U}}_N = \mathcal{U} \ominus K\mathcal{S}_N^{co}$ with $K = \begin{bmatrix} -47.5680 & -1.4807 \end{bmatrix}$ being the associated LQR, $\mathcal{S}_N = \mathcal{S}_N^{co} \oplus \mathcal{S}_N^e$.

Figure 4.11 and its associated zoom in Figure 4.12 illustrate the tube section which bounds the uncertain trajectory of the system. As the real system enters the terminal set after two instants, here only some tube sections can be seen because after time instant $k = 2$, these tubes' sections coincide. The larger zonotope is the set $\underline{\mathbf{x}}^*(\hat{\mathbf{x}}_k, k) \oplus \mathcal{S}_k$, which is in fact the tube bounding the real system. The smaller zonotope is the set $\underline{\mathbf{x}}^*(\hat{\mathbf{x}}_k, k) \oplus \mathcal{S}_k^{co}$, which bounds the error between the estimation state and the nominal state. The smallest zonotope is the guaranteed state estimation set $\hat{\mathbf{x}}_k \oplus \mathcal{S}_k^e$, which is computed based on the optimization problem [3.22]. The initial guaranteed state estimation is small because the robust invariant condition on this set must be satisfied.

The evolution of the guaranteed state estimation (zonotope $\mathcal{S}_k^e$ centered in $\hat{\mathbf{x}}_k$) is illustrated in Figure 4.13.

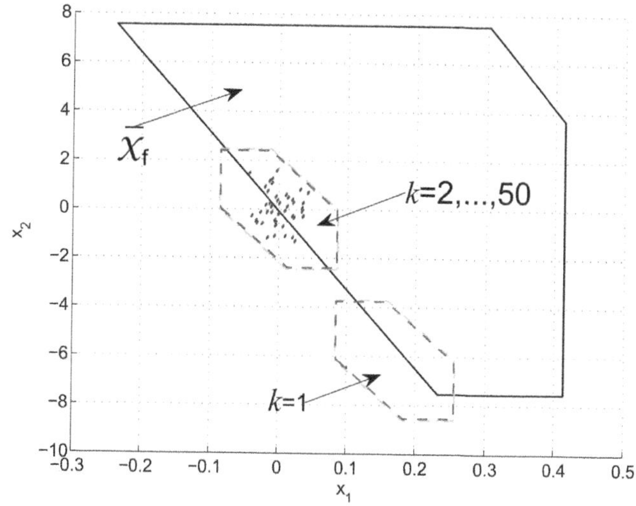

Figure 4.11. *Tube trajectory of the controlled magnetic levitation system. For a color version of this figure, see www.iste.co.uk/stoica/zonotopes.zip*

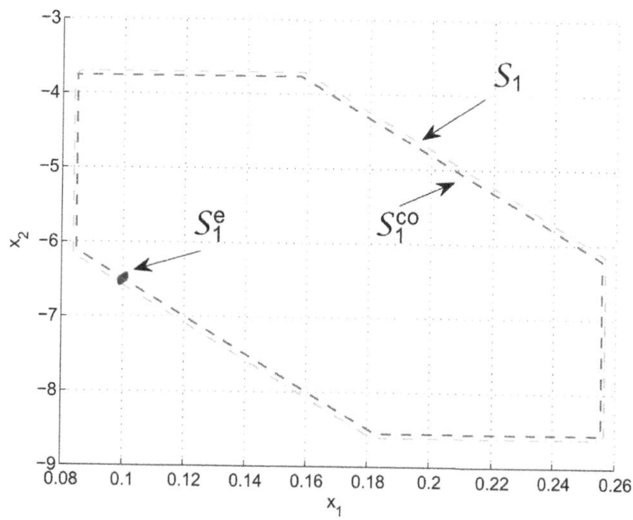

Figure 4.12. *Tube section bounding the system trajectory at the initial time instant $k = 1$. For a color version of this figure, see www.iste.co.uk/stoica/zonotopes.zip*

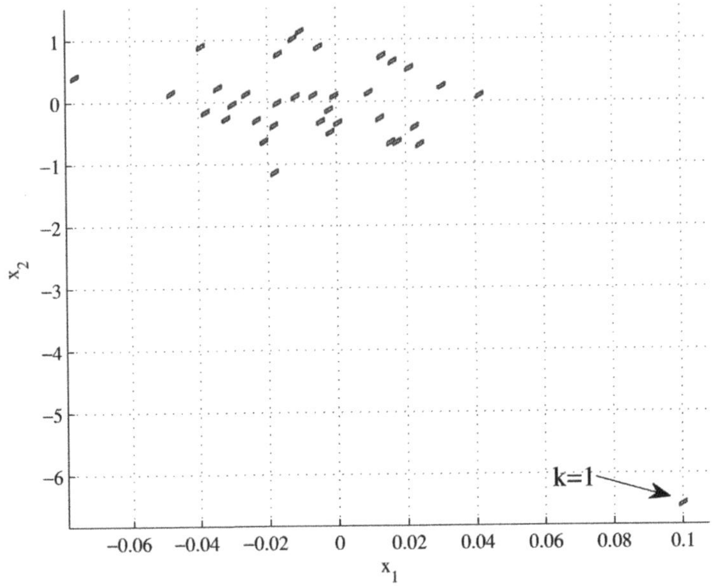

Figure 4.13. *Evolution of the guaranteed state estimation using the TMPC*

Figure 4.14 shows the stability of this output feedback system respecting the set of constraints $\mathcal{X} = \{\mathbf{x} \in \mathbb{R}^2 : x_1 \in [-0,5;0,5]\text{m}, x_2 \in [-10,10]\text{m/s}\}$. Figure 4.15 illustrates the control signal computed by the TMPC. This figure shows that the control signal respects the control constraint $u \in \mathbb{R} : |u| \leq 5\text{A}$. The pendulum position (Figure 4.16) and the pendulum speed (Figure 4.17) also respect the imposed constraints. It can be noted that the nominal state (represented in black) reaches the origin. Because of the bounded disturbances and measurement noises, the system state does not converge to the origin but it converges to a set containing the origin (S_∞ in Figure 4.14).

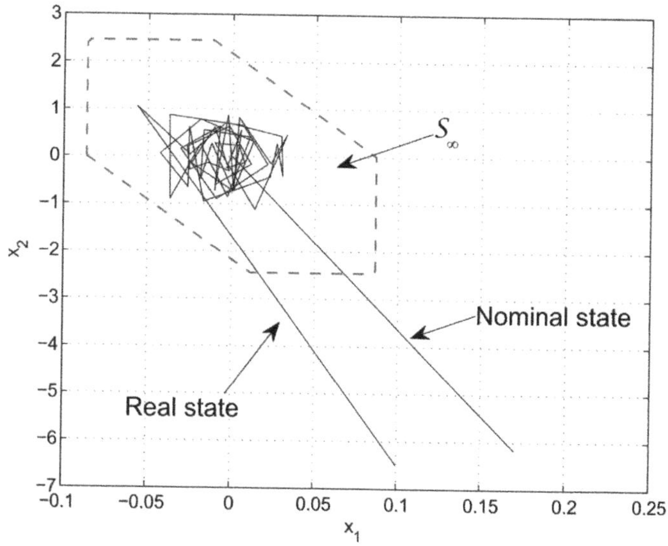

Figure 4.14. *Real state and nominal state of the closed-loop pendulum system*

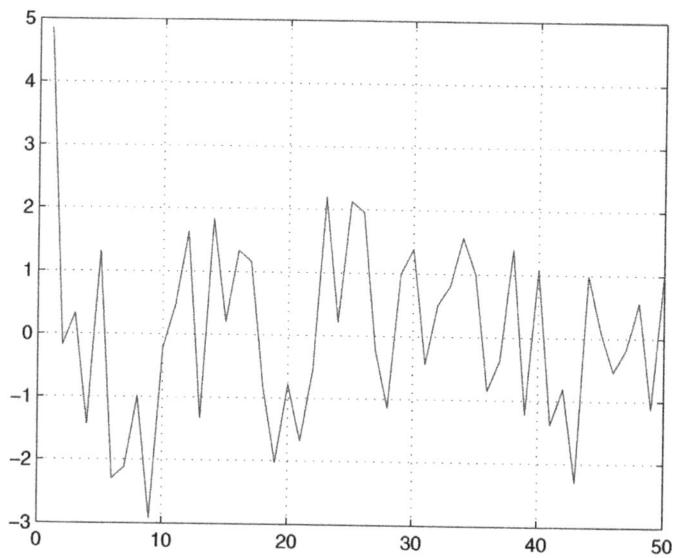

Figure 4.15. *Control signal of the closed-loop magnetic levitation system computed by the TMPC*

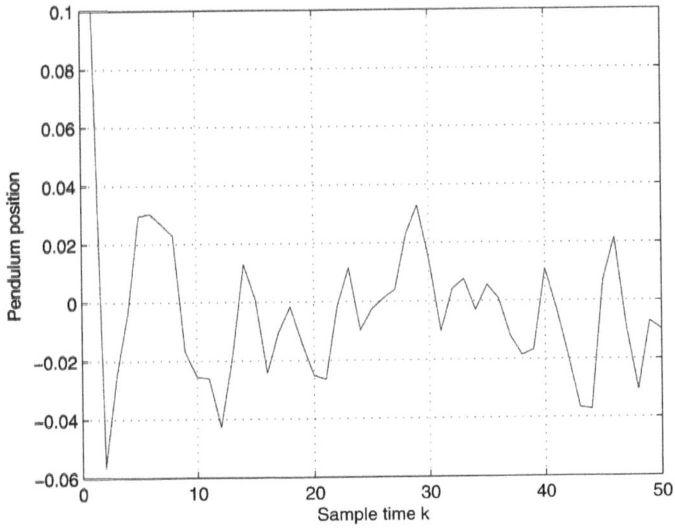

Figure 4.16. *Pendulum position controlled by the TMPC*

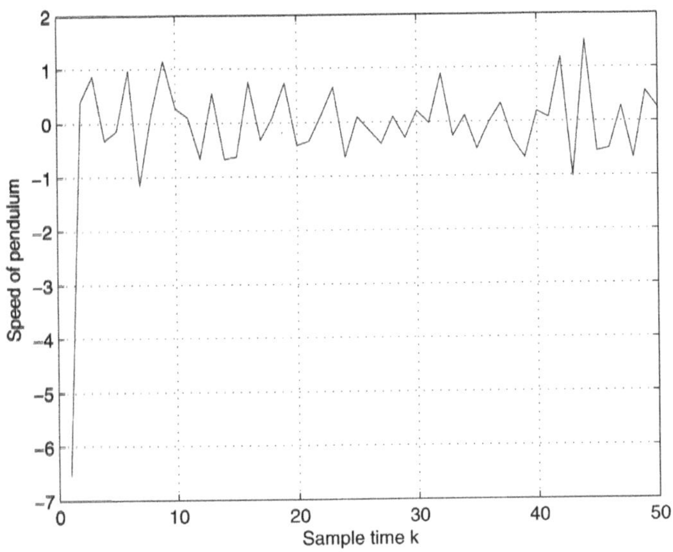

Figure 4.17. *Pendulum speed controlled by the TMPC*

Conclusion and Perspectives

This book proposes a robust predictive control technique based on zonotopic set-membership state estimation for discrete-time linear time invariant systems subject to constraints in the presence of bounded disturbances and measurement noises.

Using the deterministic approach, the disturbances and measurement noises are assumed to be unknown but bounded by zonotopes. Then a zonotopic set containing all the possible system states that are consistent with the disturbances and the noises is computed at each time instant using a set-membership estimation algorithm. An optimization criterion to compute the zonotopic guaranteed state estimation based on the minimization of the P-radius of the zonotopic estimation set is chosen as a trade-off between good accuracy of the estimation and reasonable complexity of the computation. Using the P-radius-based criterion leads to the off-line solving of a Bilinear Matrix Inequality (BMI) optimization problem [LE 11a]. To solve this problem, the *Penbmi* solver [KOČ 03] can be used. To overcome the use of BMI solvers, a suboptimal solution leading to a Linear Matrix Inequality (LMI) optimization problem [LE 11b] is addressed. First developed for Single-Output systems, the set-membership estimation procedure is then extended for the case of Multi-Output systems. The first proposed solution is based on the direct extension of the adopted solution for the case of Single-Output systems. This solution is shown to be conservative due to neglecting the coupling effect of Multi-Output systems. When the coupling effect is taken into account, the optimization problem becomes a Polynomial Matrix Inequality (PMI) problem, which is non-convex and thus difficult to solve. As a result, the proposed solution is to use a relaxation technique [HEN 06] to transform this Polynomial Matrix Inequality problem into a LMI problem. But this relaxation technique leads to an important increase in the number of scalar decision variables and, thus, limits the application of the proposed approach to large-scale systems. Moreover, the simulation result shows a poor estimation performance due to the use of this relaxation technique. In order to improve the overall performance for Multi-Output systems, this book considers the zonotopic outer approximation of the intersection between a zonotope and a polytope [LE 12b]. Using this result and the P-radius-based optimization criterion leads to a significant improvement in both the estimation performance and the complexity of the computation in comparison with the previous proposed methods (the direct extension of Single-Output case and the PMI-based method). The comparative results are analyzed in different examples in the book.

The problem of robust control for constrained uncertain systems in the predictive control context is also addressed, more precisely a Tube-based Model Predictive Control (TMPC) for systems with bounded disturbances and measurement noises which build a tube trajectory of the real system around the nominal model. This control technique allows us to guarantee the stability of the controlled system. When the system state is not available, the TMPC is usually associated with the Luenberger observer to obtain a state estimation [MAY 09]. This book focuses on the TMPC in the context of the zonotopic set-membership estimation [LE 11b], which allows us to improve the estimation performance, and thus the performance of the controlled system. Finally, the application of the proposed TMPC to the magnetic levitation system shows the effectiveness of the developed technique [LE 12a].

Several directions are proposed for future developments of this book. Applying the P-radius-based zonotopic set-membership estimation for Multi-Output linear discrete-time systems subject to bounded disturbances and measurement noises leads us to solve a PMI optimization problem. Because of the poor performance obtained by the relaxation LMI technique [HEN 06] (outer approximation of the solution of the PMI optimization problem) applied in this book, the first perspective consists of comparison of the results obtained with the relaxation procedure [HEN 06] with other existing techniques [HOL 04, KOJ 03, CHE 00] or the most recent work [HEN 11], which offers an inner approximation of the solution of the PMI optimization problem. If the reader is interested in these results, another direction to be explored consists of developing new relaxation techniques that will lead to other suboptimal solutions of the initial PMI problem in order to obtain a better performance of the estimation.

An interesting idea to pursue is the extension of the zonotopic set-membership estimation for systems with time delay and especially for systems with variable time delay [LOM 09]. If the delay can be taken into account in the mathematical model as an interval parametric uncertainty [STO 09], then the extension of this estimation technique should not be too complicated.

In the case of Multi-Output systems, the set-membership estimation is obtained using a multisensor system. Thus, a natural direction in which to explore further is the fault diagnosis problem of the multisensor system [STO 10a, PUI 10, TOR 12]. This problem can be considered using the proposed zonotopic outer approximation of the intersection between a zonotope and a polytope. If this intersection is zero, it means that a fault is detected. Then a continuing study which can be considered is the fault tolerant control [STO 10b, SER 11, YET 11].

A different perspective is to build an efficient output feedback control law (offering stability guarantees) which allows us to control linear time invariant systems with interval parametric uncertainties, disturbances, measurement noises and constraints. The solution of the TMPC associated with the set-membership estimation for systems with parametric uncertainties is still an open problem.

Finally, the application of the proposed work on a more complex system with the experimental validation can be also considered.

Appendix

Basic Matrix Operation Definitions

Some basic matrix operations used in this book are recalled here.

DEFINITION A1.1.– *A matrix* $M = M^T \in \mathbb{R}^{n\times n}$ *is called a* semi positive-definite matrix *(respectively* semi negative-definite matrix*), denoted* $M \succeq 0$ *(*$M \preceq 0$*), if* $\boldsymbol{z}^T M\boldsymbol{z} \geq 0$ *(*$\boldsymbol{z}^T M\boldsymbol{z} \leq 0$*) for all non-zero vectors* $\boldsymbol{z}$ *with real entries (*$\boldsymbol{z} \in \mathbb{R}^n$*).*

DEFINITION A1.2.– *A matrix* $M = M^T \in \mathbb{R}^{n\times n}$ *is called a* strictly positive-definite matrix *(respectively* strictly negative-definite matrix*), denoted* $M \succ 0$ *(*$M \prec 0$*), if* $\boldsymbol{z}^T M\boldsymbol{z} > 0$ *(*$\boldsymbol{z}^T M\boldsymbol{z} < 0$*) for all non-zero vectors* $\boldsymbol{z}$ *with real entries (*$\boldsymbol{z} \in \mathbb{R}^n$*).*

DEFINITION A1.3.– *A mathematical expression of the following form is called the* Linear Matrix Inequality (LMI)*:*

$$F(\boldsymbol{x}) = F_0 + \sum_{i=1}^{n} x_i F_i \succ 0$$

where $\boldsymbol{x} = \begin{bmatrix} x_1 & x_2 & \dots & x_n \end{bmatrix}^T \in \mathbb{R}^n$ *is the vector of decision variables and* F_i*,* $i = 0, \dots, n$ *are given symmetric matrices.*

The following two problems related to LMI are considered in this book:

1) *Feasibility problem*: Does a solution $\mathbf{x} \in \mathbb{R}^n$ exist such that the following LMI $F(\mathbf{x}) \succ 0$ is feasible?

2) *Optimization problem*: Minimize a linear cost function $\mathbf{b}^T\mathbf{x}$ subjected to the LMI constraint $F(\mathbf{x}) \succ 0$.

DEFINITION A1.4.– SCHUR COMPLEMENT [BOY 94, SCH 05a] – *Consider the following LMI:*

$$\begin{bmatrix} Q(\boldsymbol{x}) & S(\boldsymbol{x}) \\ S^T(\boldsymbol{x}) & R(\boldsymbol{x}) \end{bmatrix} \succeq 0$$

where $Q(\boldsymbol{x})$*,* $R(\boldsymbol{x})$ *are symmetric matrices and* $Q(\boldsymbol{x})$*,* $R(\boldsymbol{x})$*,* $S(\boldsymbol{x})$ *are affine on* $\boldsymbol{x}$*, then this LMI is equivalent to:*

$$\begin{cases} Q(\boldsymbol{x}) \succeq 0 \\ Q(\boldsymbol{x}) - S(\boldsymbol{x})R(\boldsymbol{x})^{-1}S^T(\boldsymbol{x}) \succeq 0 \end{cases}$$

or

$$\begin{cases} R(\boldsymbol{x}) \succeq 0 \\ R(\boldsymbol{x}) - S^T(\boldsymbol{x})Q(\boldsymbol{x})^{-1}S(\boldsymbol{x}) \succeq 0 \end{cases}$$

DEFINITION A1.5.– *A* Bilinear Matrix Inequality *(BMI) is defined by the following expression:*

$$F(\boldsymbol{x}) = F_0 + \sum_{i=1}^{n} x_i F_i + \sum_{i=1}^{n}\sum_{j=1}^{m} x_i x_j F_{ij} \succ 0$$

where the vector of decision variables $\boldsymbol{x} = \begin{bmatrix} x_1 & x_2 & \dots & x_n \end{bmatrix}^T \in \mathbb{R}^n$ *if* $n \geq m$ *or* $\boldsymbol{x} = \begin{bmatrix} x_1 & x_2 & \dots & x_m \end{bmatrix}^T \in \mathbb{R}^m$ *if* $n < m$*, and the given symmetric matrices* F_0*,* F_i*,* F_{ij}*, with* $i, j = 1, \dots, n$*.*

DEFINITION A1.6.– *A Polynomial Matrix Inequality (PMI) is defined by the following expression:*

$$F(\boldsymbol{x}) = \boldsymbol{x}F \succ 0$$

where $\boldsymbol{x} = \begin{bmatrix} 1 & x_1 & \dots x_n & x_1^2 & x_1x_2 & \dots & x_1x_n & \dots & x_n^n \end{bmatrix}^T \in \mathbb{R}^n$ *is the vector of decision variables and* $F = \begin{bmatrix} F_0 & F_1 & \dots & F_n & F_{11} & F_{12} & \dots & F_{1n} & \dots & F_{n\dots n} \end{bmatrix}$, *with* F_* *given symmetric matrices.*

Note that LMIs and BMIs are just a particular case of PMIs. The interested reader can find more details about LMIs, BMIs and PMIs in the literature [BOY 94, NES 94, BEN 01, GHA 00, SCH 05a, HEN 06].

Bibliography

[ALA 05] ALAMO T., BRAVO J.M., CAMACHO E.F., "Guaranteed state estimation by zonotopes", *Automatica*, vol. 41, pp. 1035–1043, 2005.

[ALA 07a] ALAMIR M., "Nonlinear moving horizon observers: theory and real-time implementation", *Lecture Notes in Control and Information Sciences*, vol. 363, pp. 139–179, 2007.

[ALA 07b] ALAMO T., RAMIREZ D.R., DE LA PEÑA D.M., CAMACHO E.F., "Minmax MPC using a tractable QP problem", *Automatica*, vol. 43, pp. 693–700, 2007.

[ALA 08] ALAMO T., BRAVO J.M., REDONDO M.J., CAMACHO E.F., "A set-membership state estimation algorithm based on DC programming", *Automatica*, vol. 44, no. 1, pp. 216–224, 2008.

[ALT 07] ALTHOFF M., STURSBERG O., BUSS M., "Reachability analysis of linear systems with uncertain parameters and inputs", *Proceedings of the 46th IEEE Conference on Decision and Control*, New Orleans, LA, vol. 41, pp. 726–732, 2007.

[ALT 10] ALTHOFF M., STURSBERG O., BUSS M., "Computing reachable sets of hybrid systems using a combination of zonotopes and polytopes", *Nonlinear Analysis: Hybrid Systems*, vol. 4, no. 2, pp. 233–249, 2010.

[AUG 06] AUGHENBAUGH J.M., PAREDIS C.J.J., "Why are intervals and imprecision important in engineering design?" *Proceedings of the Reliable Engineering Computing Workshop*, Savannah, GA, 2006.

[AYY 06] AYYUB B.M., KLIR G.J., *Uncertainty Modeling and Analysis in Engineering and the Sciences*, Chapman and Hall/CRC, 2006.

[BAR 99] BAREQUET G., HAR-PELED S., "Efficiently approximating the minimum-volume bounding box of a point set in three dimensions", *Journal of Algorithms*, vol. 38, pp. 91–109, 1999.

[BAT 08] BATABYAL A.A., *Dynamic and Stochastic Approaches to the Environment and Economic Development*, World Scientific Publishing Company, 2008.

[BEC 08] BECIS-AUBRY Y., BOUTAYEB M., DAROUACH M., "State estimation in the presence of bounded disturbances", *Automatica*, vol. 44, no. 7, pp. 1867–1873, 2008.

[BEC 11] BECIS-AUBRY Y., AUBRY D., RAMDANI N., "Multisensor set-membership state estimation of nonlinear models with potentially failing measurements", *Proceedings of the 18th World Congress IFAC*, Milan, Italy, pp. 12030–12035, 2011.

[BEC 12] BECIS-AUBRY Y., RAMDANI N., "State-bounding estimation for nonlinear models with multiple measurements", *Proceedings of the 2012 American Control Conference ACC*, Montreal, Canada, pp. 1883–1888, 2012.

[BEM 98] BEMPORAD A., "Reducing conservativeness in predictive control of constrained systems with disturbances", *Proceedings of the 37th IEEE Conference on Decision and Control*, Tampa, FL, pp. 1384–1391, 1998.

[BEM 99] BEMPORAD A., MORARI M., "Robust Model Predictive Control: a survey", *Robustness in Identification and Control*, vol. 245, pp. 207–226, 1999.

[BEM 00] BEMPORAD A., GARULLI A., "Output feedback predictive control of constrained linear systems via set-membership state estimation", *International Journal of Control*, vol. 73, no. 8, pp. 655–665, 2000.

[BEN 01] BEN-TAL A., NEMIROVSKI A., *Lectures on Modern Convex Optimization, MPSSIAM Series on Optimization*, SIAM, Philadelphia, PA, 2001.

[BER 71] BERTSEKAS D.P., RHODES I.B., "Recursive state estimation for a set-membership description of uncertainty", *IEEE Transactions on Automatic Control*, vol. 16, no. 2, pp. 117–128, 1971.

[BER 03] BERTSEKAS D.P., NEDÍC A., OZDAGLAR A.E., *Convex Analysis and Optimization*, Athena Scientific, 2003.

[BHA 02] BHATTACHARYA R., BALAS G.J., KAYA M.A., PACKARD A., "Nonlinear receding horizon control of an F-16 aircraft", *Journal Guidance Control and Dynamics*, vol. 25, no. 5, pp. 924–931, 2002.

[BLA 99] BLANCHINI F., "Set invariance in control", *Automatica*, vol. 35, no. 11, pp. 1747–1767, 1999.

[BLA 07] BLANCHINI F., MIANI S., *Set-theoretic Methods in Control*, Birkhauser, Boston, MA, 2007.

[BOY 94] BOYD S., GHAOUI L.E., FERON E., BALAKRISHNAN V., *Linear Matrix Inequalities in System and Control Theory*, SIAM, Philadelphia, PA, 1994.

[BRA 06] BRAVO J.M., ALAMO T., CAMACHO E.F., "Bounded error identification of systems with time-varying parameters", *IEEE Transactions on Automatic Control*, vol. 51, no. 7, pp. 1144–1150, 2006.

[BRO 97] BROWN R.G., HWANG P.Y.C., *Introduction to Random Signals and Applied Kalman Filtering*, John Wiley and Sons, 1997.

[BRO 08] BRONSTEIN E.M., "Approximation of convex sets by polytopes", *Journal of Mathematical Sciences*, vol. 153, no. 6, pp. 727–762, 2008.

[BUR 00] BURDEN R.L., FAIRES J.D., *Numerical Analysis*, Brooks Cole, 2000.

[CAM 87] CAMPO P.J., MORARI M., "Robust Model Predictive Control", *Proceedings of the American Control Conference*, Minneapolis, MN, pp. 1021–1026, 1987.

[CAM 04] CAMACHO E.F., BORDONS C., *Model Predictive Control*, Springer-Verlag, London, 2004.

[CHE 81] CHERNOUS'KO F.L., "Ellipsoidal estimates of a controlled system's attainability domain", *Journal of Applied Mathematics and Mechanics*, vol. 45, no. 1, pp. 7–12, 1981.

[CHE 00] CHESI G., GARULLI A., TESI A., VICINO A., "An LMI-based approach for characterizing the solution set of polynomial systems", *Proceedings of the 39th IEEE Conference Decision and Control*, Sydney, Australia, pp. 1501–1506, 2000.

[CHI 96] CHISCI L., GARULLI A., ZAPPA G., "Recursive state bounding by parallelotopes", *Automatica*, vol. 32, pp. 1049–1055, 1996.

[CHI 02] CHISCI L., ZAPPA G., "Feasibility in predictive control of constrained linear systems: the output feedback case", *International Journal of Robust and Non Linear Control*, vol. 12, pp. 465–487, 2002.

[CLA 87] CLARKE D.W., MOHTADI C., TUFFS P.S., "Generalized Predictive Control, Part I: the basic algorithm; Part II: extensions and interpretations", *Automatica*, vol. 23, no. 2, pp. 137–160, 1987.

[CLA 88] CLARKE D.W., "Application of Generalized Predictive Control to industrial processes", *IEEE Control System Magazine*, vol. 122, pp. 49–55, 1988.

[CLA 91] CLARKE D.W., SCATOLLINI R., "Constrained receding horizon predictive control", *Control Theory and Applications, Proceedings IEE-D*, vol. 138, no. 4, pp. 347–354, 1991.

[COM 03] COMBASTEL C., "A state bounding observer based on zonotopes", *Proceedings of the European Control Conference*, Cambridge, UK, 2003.

[COM 04] COMBASTEL C., "Observation de systèmes non linéaires appliqué à un modèle de bioréacteur. Une approche ensembliste basée sur les zonotopes", *Journal Européen des Systèmes Automatisés*, vol. 38, pp. 933–957, 2004.

[COM 05] COMBASTEL C., "A state bounding observer for uncertain non-linear continuous-time based on zonotopes", *Proceedings of the 44th IEEE Conference on Decision and Control, and the European Control Conference*, Seville, Spain, 2005.

[COM 08] COMBASTEL C., ZHANG Q., LALAMI A., "Fault diagnosis based on the enclosure of parameters estimated with an adaptive observer", *Proceedings of the 17th World Congress IFAC*, Seoul, Korea, pp. 7314–7319, 2008.

[CUT 80] CUTLER C.R., RAMAKER B.C., "Dynamic matrix control: a computer control algorithm", *Proceedings of the Automatic Control Conference*, San Francisco, CA, 1980.

[DAN 72] DANTZIG G.B., Fourier-Motzkin elimination and its dual, Report, DTIC document, 1972.

[DUR 01] DURIEU C., WALTER E., POLYAK B., "Multi-input multi-output ellipsoidal state bounding", *Journal of Optimization Theory and Applications*, vol. 111, no. 2, pp. 273–303, 2001.

[FOO 08] FOO Y.K., SOH Y.C., "Robust Kalman filtering for uncertain discrete-time systems with probabilistic parameters bounded within a polytope", *Systems and Control Letters*, vol. 57, no. 6, pp. 482–488, 2008.

[FUK 99] FUKUDA K., *Cdd/Cdd+ Reference Manual*, Institute for Operations Research, Zurich, Switzerland, 1999.

[FUK 04] FUKUDA K., "From the zonotope construction to the Minkowski addition of convex polytopes", *Journal of Symbolic Computation*, vol. 38, no. 4, pp. 1261–1272, 2004.

[GAR 89] GARCÍA C.E., PRETT D.M., MORARI M., "Model Predictive Control: theory and practice – a survey", *Automatica*, vol. 25, no. 3, pp. 335–348, 1989.

[GAR 01] GARULLI A., VICINO A., "Set membership localization of mobile robots via angle measurement", *IEEE Transactions on Robotic and Automation*, vol. 17, no. 4, pp. 450–463, 2001.

[GHA 99] GHAOUI L.E., CALAFIORE G., "Worst-case state prediction under structured uncertainty", *Proceedings of the American Control Conference*, San Diego, CA, pp. 3402–3406, 1999.

[GHA 00] GHAOUI L.E., NICULESCU S.-I., *Advances in Linear Matrix Inequality Methods in Control*, SIAM, Philadelphia, PA, 2000.

[GIL 91] GILBERT E.G., TAN K.T., "Linear systems with state and control constraints: the theory and application of maximal output admissible sets", *IEEE Transactions on Automatic Control*, vol. 36, no. 9, pp. 1008–1020, 1991.

[GIR 05] GIRARD A., "Reachability of uncertain linear systems using zonotopes", *Hybrid Systems: Computation and Control, vol. 3414 of Lecture Notes in Computer Science*, Springer, pp. 291–305, 2005.

[GOO 04] GOODWIN G., SERON M.M., De DONÁ J.A., *Constrained Control and Estimation: an Optimisation Approach*, Springer, 2004.

[GRE 84] GRECO C., MENGA G., MOSCA E., ZAPPA G., "Performance improvement of selftuning controllers by multistep horizons: the MUSMAR approach", *Automatica*, vol. 20, pp. 681–100, 1984.

[GRI 90] GRIZZLE J.W., MORAAL P.E., "Observer based control of nonlinear discrete-time systems", *Proceedings of the 29th IEEE Conference on Decision and Control*, Honolulu, HI, pp. 760–767, 1990.

[GRI 93] GRITZMANN P., STURMFELS B., "Minkowski addition of polytopes: computational complexity and applications to Gröbner bases", *SIAM Journal on Discrete Mathematic*, vol. 6, no. 2, pp. 246–269, 1993.

[GUE 08] GUERRA P., PUIG V., WITCZAK M., "Robust fault detection with unknown-input interval observers using zonotopes", *Proceedings of the 17th World Congress IFAC*, Seoul, Korea, pp. 5557–5562, 2008.

[GUI 05] GUIBAS L.J., NGUYEN A., ZHANG L., "Zonotopes as bounding volume", *Proceedings of the Symposium on Discrete Algorithm*, pp. 803–812, 2005.

[HAN 65] HANSEN E.R., "Interval arithmetic in matrix computations", *SIAM Journal on Numerical Analysis: Series B*, vol. 2, no. 2, pp. 308–320, 1965.

[HEN 02] HENRION D., LASSERRE J.B., "GloptiPoly: global optimization over polynomials with Matlab and SeDuMi", *ACM Transactions on Mathematical Software*, vol. 29, pp. 165–194, 2002.

[HEN 06] HENRION D., LASSERRE J.B., "Convergent relaxations of Polynomial Matrix Inequalities and static output feedback", *IEEE Transactions on Automatic Control*, vol. 51, no. 2, pp. 192–202, 2006.

[HEN 11] HENRION D., LASSERRE J.B., Inner approximations for Polynomial Matrix Inequalities and robust stability regions, Report, LAAS-report 11210, 2011.

[HOL 04] HOL C.W.J., SCHERER C.W., "Sum of Squares relaxations for polynomial semidefinite programming", *Proceedings of the Symposium on Mathematical Theory of Networks and Systems*, Leuven, Belgium, 2004.

[HUN 93] HUNTTENLOCHER D.P., KLANDERMAN G.A., RUCKLIDGE W.J., "Comparing images using the Hausdorff distance", *IEEE Transactions on Pattern Analysis and Machine Intelligence*, vol. 15, no. 9, pp. 850–863, 1993.

[ING 08] INGIMUNDARSON A., BRAVO J.M., PUIG V., ALAMO T., GUERRA P., "Robust fault detection using zonotope-based set-membership consistency", *International Journal of Adaptive Control and Signal Processing*, vol. 23, no. 4, pp. 311–330, 2008.

[JAU 01] JAULIN L., KIEFFER M., DIDRIT O., WALTER E., *Applied Interval Analysis*, Springer, 2001.

[KAL 60] KALMAN R.E., "A new approach to linear filtering and prediction problems", *Transactions of the ASME–Journal of Basic Engineering*, vol. 82, no. Series D, pp. 35–45, 1960.

[KEE 88] KEERTHI S.S., GILBERT E.G., "Optimal infinite-horizon feedback laws for a general class of constrained discrete-time systems: stability and moving-horizon approximations", *Journal of Optimization Theory and Applications*, vol. 57, no. 2, pp. 265–293, 1988.

[KEY 85] DE KEYSER R.M.C., VAN CAUWENBERGHE A.R., "Extended prediction self-adapted control", *IFAC Symposium on Identification and System Parameter Estimation*, York, UK, pp. 1317–1322, 1985.

[KIE 02] KIEFFER M., JAULIN L., WALTER E., "Guaranteed recursive nonlinear state estimation using interval analysis", *International Journal of Adaptive Control and Signal Processing*, 2002.

[KOČ 03] KOČVARA M., STINGL S., "PENNON A code for convex nonlinear and semidefinite programming", *Optimization Methods and Software*, vol. 18, no. 3, pp. 317–333, 2003.

[KOJ 03] KOJIMA M., Sums of squares relaxations of polynomial semidefinite programs, Report, Department of Mathematical and Computing Sciences, Tokyo Institute of Technology, 2003.

[KOT 96] KOTHARE M.V., BALAKRISHNAN V., MORARI M., "Robust constrained Model Predictive Control using Linear Matrix Inequalities", *Automatica*, vol. 32, no. 10, pp. 1361–1379, 1996.

[KÜH 98a] KÜHN W., "Rigorously computed orbits of dynamical systems without the wrapping effect", *Computing*, vol. 61, pp. 47–67, 1998.

[KÜH 98b] KÜHN W., "Toward an optimal control of wrapping effect", *International Symposium on Scientific Computing, Computer Arithmetic and Validated Numerics*, pp. 125–134, 1998.

[KUR 96] KURZHANSKI A.B., VÁLYI I., *Ellipsoidal Calculus for Estimation and Control*, Birkhauser, Boston, MA, 1996.

[KUR 06] KURZHANSKIY A.A., VARAIYA P., "Ellipsoidal toolbox", Report no. UCB/EECS-2006–46, EECS Department, University of California, Berkeley, May 2006.

[KVA 04] KVASNICA M., GRIEDER P., BAOTIĆ M., "Multi-Parametric Toolbox (MPT)", 2004.

[LAL 08] LALAMI A., Diagonostic et approches ensemblistes à base des zonotopes, PhD Thesis, University of Cergy-Pontoise, 2008.

[LAN 04] LANGSON W., CHRYSSOCHOOS I., RAKOVIĆ S., MAYNE D., "Robust Model Predictive Control using tubes", *Automatica*, vol. 40, no. 1, pp. 125–133, 2004.

[LAY 82] LAY S.R., *Convex Sets and their Applications*, Wiley, New York, 1982.

[LE 11a] LE V.T.H., ALAMO T., CAMACHO E.F., STOICA C., DUMUR D., "A new approach for guaranteed state estimation by zonotopes", *Proceedings of the 18th World Congress IFAC*, Milan, Italy, pp. 9242–9247, 2011.

[LE 11b] LE V T.H., STOICA C., DUMUR D., ALAMO T., CAMACHO E.F., "Robust tube-based constrained predictive control via zonotopic set-membership estimation", *Proceedings of the 50th IEEE Conference on Decision and Control and European Control Conference*, Orlando, FL, pp. 4580–4585, 2011.

[LE 12a] LE V.T.H., STOICA C., DUMUR D., ALAMO T., CAMACHO E.F., "Commande prédictive robuste par des techniques d'observateurs basées sur des ensembles zonotopiques", *Journal Européen des Systèmes Automatisés*, vols. 2–3, pp. 235–250, 2012.

[LE 12b] LE V.T.H., STOICA C., DUMUR D., ALAMO T., CAMACHO E.F., "Guaranteed state estimation by zonotopes for systems with interval uncertainties", *Small Workshop on Interval Methods*, Oldenburg, Germany, 2012.

[LE 12c] LE V.T.H., Commande prédictive robuste par des techniques d'observateurs à base d'ensembles zonotopiques, PhD Thesis, Ecole Supérieure d'Electricité (Supélec), Gif sur Yvette, 2012.

[LE 13] LE V.T.H., STOICA C., ALAMO T., CAMACHO E.F., DUMUR D., "Zonotope-based set-membership estimation for Multi-Output uncertain systems", *IEEE Multi-Conference on Systems and Control*, Hyderabad, India, 2013.

[LEE 97] LEE J.H., YU Z., "Worst-case formulations of Model Predictive Control for systems with bounded parameters", *Automatica*, vol. 33, no. 5, pp. 763–781, 1997.

[LEM 85] LEMOS J.M., MOSCA E., "A multipredictor-based LQ self-tuning controller", *IFAC Symposium on Identification and System Parameter Estimation*, York, UK, pp. 137–141, 1985.

[LIM 02] LIMON D., ALAMO T., CAMACHO E.F., "Stability analysis of systems with bounded additive uncertainties based on invariant sets: stability and feasibility of MPC", *Proceedings of the American Control Conference*, Anchorage, AK, pp. 364–369, 2002.

[LIM 07] LIMON D., ALAMO T., BRAVO J.M., *et al.*, *Interval Arithmetic in Robust Nonlinear MPC*, Springer-Verlag, pp. 317–326, 2007.

[LOM 09] LOMBARDI W., OLARU S., NICULESCU S.I., "Invariant sets for a class of linear systems with variable time-delay", *Proceedings of the European Control Conference*, Budapest, Hungary, 2009.

[LUE 64] LUENBERGER D.G., "Observing the state of a linear system", *IEEE Transactions on Military Electronics*, vol. 8, pp. 74–80, 1964.

[LUE 65] LUENBERGER D.G., "Observers for multivariable systems", *IEEE Transactions on Automatic Control*, vol. 11, pp. 190–197, 1965.

[MAC 00] MACIEJOWSKI J.M., *Predictive Control: A Unified Approach*, Prentice-Hall, 2000.

[MAR 06] MARE J.B., DE DONA J.A., "Moving horizon estimation of constrained nonlinear systems by Carleman approximations", *Proceedings of the 45th IEEE Conference on Decision and Control*, San Diego, CA, pp. 2147–2152, 2006.

[MAY 79] MAYBECK P.S., *Stochastic Models, Estimation and Control*, Academic Press, 1979.

[MAY 00] MAYNE D.Q., RAWLINGS J.B., RAO C.V., SCOKAERT P.O.M., "Constrained Model Predictive Control: stability and optimality", *Automatica*, vol. 36, pp. 789–814, 2000.

[MAY 05] MAYNE D.Q., SERON M.M., RAKOVIĆ S.V., "Robust Model Predictive Control of constrained linear system with bounded disturbances", *Automatica*, vol. 41, pp. 219–224, 2005.

[MAY 09] MAYNE D.Q., RAKOVIĆ S.V., FINDEISEN R., ALLGÖWER F., "Robust output feedback Model Predictive Control of constrained linear system: time varying case", *Automatica*, vol. 45, pp. 2082–2087, 2009.

[MES 10] MESLEM N., RAMDANI N., CANDAU Y., "Using hybrid automata for set-membership state estimation with uncertain nonlinear continuous-time systems", *Journal of Process Control*, vol. 20, no. 4, pp. 481–489, 2010.

[MIC 91] MICHALSKA H., MAYNE D.Q., "Moving horizon observers", *Proceedings of the IFAC Symposium Nonlinear Control System Design*, Bordeaux, France, pp. 576–581, 1991.

[MIL 96] MILANESE M., NORTON J., PIET-LANANIER S., WALTER E., *Bounding Approach to System Identification*, Plenum Press, 1996.

[MON 89] MONTGOMERY H., "Computing the volume of a zonotope", *The American Mathematicals Monthly*, vol. 97, p. 431, 1989.

[MOO 66] MOORE R.E., *Interval Analysis*, Prentice-Hall, Englewood Cliff, NJ, 1966.

[MOO 75] MOORE J.B., LEDWICH G.F., "Minimal order observers for estimating linear functions of a state vector", *IEEE Transactions on Automatic Control*, vol. 20, no. 5, pp. 623–632, 1975.

[MOR 94] MORARI M., *Advances in Model-Based Predictive Control*, Oxford University Press, 1994.

[MUR 73] MURDOCH P., "Observer design for a linear functional of the state vector", *IEEE Transactions on Automatic Control*, vol. 18, no. 3, pp. 308–310, 1973.

[NES 94] NESTEROV Y., NEMIROVSKI A., *Interior Point Polynomial Methods in Convex Programming: Theory and Applications*, SIAM, Philadelphia, PA, 1994.

[OHR 95] OHRN K., AHLEN A., STERNARD M., "A probabilistic approach to multivariable robust filtering and open-loop control", *IEEE Transactions on Automatic Control*, vol. 40, no. 3, pp. 405–418, 1995.

[OLA 05] OLARU S., Contribution à l'étude de la commande prédictive sous contraintes par approche géométrique, PhD Thesis, University of Paris Sud-Supélec, 2005.

[POL 04] POLYAK B.T., NAZIN S.A., DURIEU C., WALTER E., "Ellipsoidal parameter or state estimation under model uncertainty", *Automatica*, vol. 40, pp. 1171–1179, 2004.

[PUI 01] PUIG V., CUGUERÓ P., QUEVEDO J., "Worst-case estimation and simulation of uncertain discrete-time systems using zonotopes", *Proceedings of the European Control Conference*, Portugal, 2001.

[PUI 03] PUIG V., SALUDES J., QUEVEDO J., "Worst-case simulation of discrete linear time-invariant interval dynamic systems", *Reliable Computing*, vol. 9, no. 4, pp. 251–290, 2003.

[PUI 10] PUIG V., "Fault diagnosis and fault tolerant control using set-membership approaches: application to real case studies", *Applied Mathematics and Computer Science*, vol. 20, no. 4, pp. 619–635, 2010.

[QIN 03] QIN S.J., BADGWELL T.A., "A survey of industrial Model Predictive Control technology", *Control Engineering Practice*, vol. 11, pp. 733–764, 2003.

[RAÏ 04] RAÏSSI T., RAMDANI N., CANDAU Y., "Set membership state and parameter estimation for systems described by nonlinear differential equations", *Automatica*, vol. 40, no. 10, pp. 1771–1777, 2004.

[RAK 05] RAKOVIĆ S.V., KERRIGAN E.C., KOURAMAS K.I., MAYNE D.Q., "Invariant approximation of the minimal robustly positively invariant set", *IEEE Transactions on Automatic Control*, vol. 50, no. 3, pp. 406–410, 2005.

[RIC 78] RICHALET J., RAULT A., TESTUD J.L., PAPON J., "Model Predictive Heuristic Control: application to industrial processes", *Automatica*, vol. 14, no. 5, pp. 413–428, 1978.

[ROC 70] ROCKAFELLAR R.T., *Convex analysis*, Princeton University Press, 1970.

[ROS 03] ROSSITER J.A., *Model-Based Predictive Control: A practical approach*, CRC Press LLC, 2003.

[SCH 68] SCHWEPPE F.C., "Recursive state estimation: unknown but bounded errors and system inputs", *IEEE Transactions on Automatic Control*, vol. 13, no. 1, pp. 22–28, 1968.

[SCH 05a] SCHERER C.S., WEILAND S., "Linear Matrix Inequalities in control", 2005.

[SCH 05b] SCHÖN S., KUTTERER H., "Using zonotopes for overestimation-free interval least-squares: some geodetic applications", *Reliable Computing Springer*, vol. 11, pp. 137–155, 2005.

[SCI 11] SCIBILIA F., OLARU S., HOVD M., "On feasible sets for MPC and their approximations", *Automatica*, vol. 47, no. 1, pp. 133–139, 2011.

[SCO 95] SCOKAERT P.O.M., RAWLINGS J.B., "Stability of Model Predictive Control under perturbations", *Proceedings of the IFAC Symposium on Nonlinear Control Systems Design*, Lake Tahoe, CA, pp. 1317–1322, 1995.

[SCO 99] SCOKAERT P.O.M., MAYNE D.Q., RAWLINGS J.B., "Suboptimal Model Predictive Control (feasibility implies stability)", *IEEE Transactions on Automatic Control*, vol. 44, no. 3, pp. 648–654, 1999.

[SER 11] SERON M.M., De DONÁ J.A., RICHTER J., "Fault tolerant control using virtual actuators and set-separation detection principles", *International Journal of Robust and Nonlinear Control*, vol. 22, pp. 709–742, 2011.

[SEY 94] SEYMOUR P., "A note on hyperplane generation", *Journal of Combinatorial Theory, Series B*, vol. 61, no. 1, pp. 88–91, 1994.

[SHE 74] SHEPHARD G., "Combinatorial properties of associated zonotopes", *Canadian Journal of Mathematics*, vol. 26, pp. 302–321, 1974.

[SOE 92] SOETERBOEK R., *Predictive Control: A Unified Approach*, Prentice-Hall, 1992.

[STE 95] STERNAD M., OHRN K., AHLEN A., "Robust H_2 filtering for structured uncertainty: the performance of probabilistic and minimax schemes", *Proceedings of the European Control Conference*, Rome, Italy, pp. 87–92, 1995.

[STO 08] STOICA C., Robustification de lois de commande prédictives multivariables, PhD Thesis, University of Paris Sud-Supélec, 2008.

[STO 09] STOICA C., ARAHAL M.R., RIVERA D.E., *et al.*, "Application of robustified Model Predictive Control to a production-inventory system", *Proceedings of the 48th IEEE Conference on Decision and Control and 28th Chinese Control Conference*, Shanghai, China, pp. 3993–3998, 2009.

[STO 10a] STOICAN F., OLARU S., BITSORIS G., "A fault detection scheme based on controlled invariant sets for multi-sensor systems", *Proceedings of the 2010 Conference on Control and Fault Tolerant Systems*, Nice, France, pp. 468–473, 2010.

[STO 10b] STOICAN F., OLARU S., SERON M.M., DE DONÁ J.A., "A fault tolerant control scheme based on sensor switching and dwell time", *Proceedings of the 49th IEEE Conference on Decision and Control*, Atlanta, GA, pp. 756–761, 2010.

[STO 11] STOICAN F., Fault tolerant control based on set-theoretic methods, PhD Thesis, University of Paris Sud-Supélec, 2011.

[STR 05] STRANG G., *Linear Algebra and its Applications*, Brooks Cole, 2005.

[STU 03] STURSBERG O., KROGH B., "Efficient representation and computation of reachable sets for hybrid systems", *Hybrid Systems: Computation and Control, vol. 2623 of Lecture Notes in Computer Science*, Springer, pp. 482–497, 2003.

[TOR 12] TORNIL-SIN S., OCAMPO-MARTINEZ C., PUIG V., ESCOBET T., "Robust fault detection of non-linear systems using set-membership state estimation based on constraint satisfaction", *Engineering Applications of Artificial Intelligence*, vol. 25, no. 1, pp. 1–10, 2012.

[ULL 11] ULLAH M., WOLKENHAUER O., *Stochastic Approaches for Systems Biology*, Springer, 2011.

[VIC 96] VICINO A., ZAPPA G., "Sequential approximation of feasible parameter sets for identification with set-membership uncertainty", *IEEE Transactions on Automatic Control*, vol. 41, pp. 774–785, 1996.

[WAL 89] WALTER E., PIET-LAHANIER H., "Exact recursive polyhedral description of the feasible parameter set for bounded-error models", *IEEE Transactions on Automatic Control*, vol. 34, no. 8, pp. 911–915, 1989.

[WAN 01] WAN Z., KOTHARE M.V., "Robust output feedback model predictive control using off-line Linear Matrix Inequalities", *Journal of Process Control*, vol. 12, no. 7, pp. 763–774, 2001.

[WIT 68] WITSENHAUSEN S.H., "Sets of possible states of linear systems given perturbed observations", *IEEE Transactions on Automatic Control*, vol. 13, pp. 556–558, 1968.

[XIE 94] XIE L., SOH Y.C., DE SOUZA C.E., "Robust Kalman filtering for uncertain discrete time systems", *IEEE Transactions on Automatic Control*, vol. 39, no. 6, pp. 1310–1314, 1994.

[YDS 84] YDSTIE B.E., "Extended horizon adaptive control", *Proceedings of the 9th World Congress IFAC*, Budapest, Hungary, 1984.

[YET 11] YETENDJE A., De DONÁ J.A., SERON M.M., "Multisensor fusion fault tolerant control", *Automatica*, vol. 47, no. 7, pp. 1461–1466, 2011.

[ZIE 95] ZIEGLER G.M., *Lecture on Polytopes*, Springer, 1995.

Index

Printed and bound by CPI Group (UK) Ltd, Croydon, CR0 4YY
16/09/2026
14963183-0002